JOURNEYMAN ELECTRICIAN

EXAM PREP STUDY GUIDE

RODNEY REED

PrepWise
PUBLISHING

This page is intentionally left blank

Table of Contents

Preface

I was 23 when I was an apprentice, and my mentor was a gentle man named Frank Bowman. Everyone called him "Steady Frank" because he never rushed a job, no matter how much the foreman pushed. I remember my first big commercial project—a three-story office building with what felt like miles of conduit to run.

Frank watched me struggling to make a perfect bend on a piece of 1-inch EMT. After my third attempt didn't quite work out, I set my tools down with a sigh. Frank just chuckled, picked up the bender, and said, "Son, electricity follows the path you give it. It doesn't care if it takes you three tries or thirty. But your wallet sure will."

He showed me his hands—scarred and calloused from thirty years in the trade—and added, "These marks tell a story. Each one taught me something. Learn now with me watching so you'll be ready when you're alone on a job with the power company waiting to flip the switch."

That lesson stuck with me through my entire career. Frank taught me that becoming an electrician goes beyond memorizing code articles or knowing how to wire a three-way switch. It's about developing judgment that keeps you and others safe when you're facing a puzzle of wires that all look similar at first glance.

I still remember the day I was preparing for my journeyman exam. Frank dropped by my apartment, uninvited, with a stack of dog-eared code books and a thermos of coffee. "Thought you might need some company," he said, pulling up a chair at my kitchen table. We stayed up until 2 AM, with Frank firing questions at me about load calculations, box fill, and grounding requirements.

When I passed my exam, Frank was the first person I called. His response? "Well, I would've been mighty surprised if you hadn't." That was Frank—he gave compliments in his own way, and his faith in me spoke volumes.

Now, twenty years later, I am in Frank's position and I've trained dozens of apprentices, many who built lasting careers in the trade, others who discovered different callings. But I've always tried to pass on what Frank taught me: patience, precision, and pride in your work.

The journeyman electrician exam marks a milestone in your career. It represents your growth from someone who follows instructions to someone who makes decisions—important ones that affect people's safety and livelihoods. This test offers you the chance to prove to yourself that you've mastered the basics of a complex and rewarding craft.

So as you flip through these pages and tackle these practice questions, remember that you're preparing to join the ranks of professionals who keep the lights on, the motors running, and the world powered up. It's a responsibility worth embracing, and a skill set worth mastering completely.

1

Mastering the Exam Format

When I walked into the testing center for my journeyman exam, there was a rush of adrenaline that I felt. There were rows and rows of tables filled with focused electricians, each working through thick code books and calculators. Some had color-coded tabs sticking out from their NEC manuals like porcupine quills. Others were making quick notes. The proctor at the front looked like he'd seen it all before—hundreds of electricians transforming from test-takers to certified journeymen right before his eyes.

Frank had prepared me for this moment. "The test checks if you know the rules well enough to work independently," he told me. "It's your chance to prove you can handle the responsibility that comes with the license." That advice changed how I studied, and it's why I'm sharing it with you now. Understanding exactly what you're facing will boost your confidence and set you up for success on exam day.

This exam is essentially your introduction to a new chapter in your career. It's not just about passing—it's about proving to yourself that you've mastered the fundamentals of a trade that powers our modern world. The questions might seem tough, but remember: every licensed journeyman in the country sat where you're sitting now, and they made it through. You will too.

I've never regretted the hours spent studying or the practice tests I took since I passed the exam. That preparation built the foundation for everything that came after—from running my first solo job to training apprentices of my own. Your journey through this exam matters, and the habits you build while studying will serve you throughout your career.

1.1 Understanding the Structure of the Exam

The first step to mastering the exam is to understand its layout. Knowing how the exam is structured will help you approach your study plan strategically and feel more at ease when test day arrives.

The Format of the Exam

Most journeyman electrician exams follow a similar pattern across the country, though specific details may vary by state. Here's what you can typically expect:

The standard journeyman exam contains between 70-100 multiple-choice questions, with 80 questions being the most common format. You'll need to answer these questions within a 4-hour time frame, giving you about 3 minutes per question on average. Some states split the exam into two parts, while others administer it in a single sitting.

To pass, you'll need to score at least 70% in most jurisdictions, though some states set the bar at 75%. That means if you're facing an 80-question test, you'll need to get at least 56-60 questions right.

The exam breaks down into several key sections:

- **General electrical knowledge** (approximately 10% of the exam): Covers basic electrical theory, Ohm's Law, circuit calculations, and terminology.
- **NEC knowledge and application** (approximately 50-60% of the exam): The meat of the test focuses on your ability to navigate and apply the National Electrical Code correctly.
- **Wiring methods and materials** (approximately 15% of the exam): Tests your knowledge of proper installation practices, raceway systems, and conductor selection.
- **Equipment installation** (approximately 10% of the exam): Covers motors, transformers, and service equipment installation requirements.
- **Safety practices** (approximately 10% of the exam): Addresses OSHA requirements, lockout/tagout procedures, and job site safety protocols.
- **Plan reading and calculations** (approximately 5-10% of the exam): Tests your ability to interpret electrical plans and perform load calculations.

Table 1 below depicts several key sections of the exam, their approximate percentage of contribution, and the descriptions.

Topic	Approximate % of Exam Content	Description
General electrical knowledge	10%	Covers basic electrical theory, Ohm's Law, circuit calculations, and terminology
NEC knowledge and application	50-60%	The meat of the test focuses on your ability to navigate and apply the National Electrical Code correctly.
Calculations	15-20%	Tests your ability to perform load calculations, sizing conductors, voltage drop, and other mathematical aspects of electrical work.
Plan reading & diagrams	10-15%	Assesses your ability to interpret blueprints, schematics, and diagrams.
Safety & troubleshooting	5-10%	Covers OSHA requirements, safe work practices, and diagnosing common electrical problems.

Table 1: Journeyman Exam Weighting Table

Each question on the test falls into one of these categories, though they won't be labeled as such on the exam. The mix of questions aims to test both your theoretical knowledge and practical application skills.

My experience has shown that most test-takers find the code-related questions the most challenging, not because the material is difficult, but because finding the right information quickly takes practice. The exam is open-book, meaning you can use your NEC during the test, but without proper preparation, you might spend too much time hunting for answers.

Passing Score Requirements

Most states set the passing threshold at 70%, though some jurisdictions require 75%. On an 80-question exam, this means you'll need to answer 56-60 questions correctly to pass.

Every question is scored as either correct (1 point) or incorrect (0 points). There's no partial credit for "almost right" answers, and no penalty for wrong answers. This scoring approach drives an important test strategy: never leave a question blank, even if you must guess.

The exam doesn't tell you which questions you got right or wrong after completion. You'll typically receive only your overall score or a pass/fail notification, sometimes with a breakdown by major content area.

1.2 Exam-Day Expectations and Best Practices

The day of the exam is the culmination of all your preparation. Understanding what to expect and how to behave on exam day can make all the difference. This section covers strategies for staying calm, prepared, and focused.

What to Expect on Exam Day

The room was quieter than a church on Tuesday, and strangely enough, I wasn't fully prepared for the formal testing environment, but let me save you from that surprise. Most journeyman exams take place at designated testing centers, state licensing facilities, or community college classrooms. These locations are set up to minimize distractions and prevent cheating. Tables are typically spaced well apart, and the room is kept quiet throughout the test.

You'll need to bring several items with you:

- Government-issued photo ID (driver's license or passport)
- Confirmation letter or email showing your registration
- Your NEC codebook (most states allow this, but check your jurisdiction's rules)
- Calculator (usually only basic or non-programmable calculators are permitted)
- #2 pencils and erasers
- A watch to keep track of time (as long as it doesn't have smart features)

What you cannot bring often includes:

- Cell phones or smart watches
- Reference materials beyond the approved NEC book
- Notes or writing on anything except your test materials
- Food (unless you have medical documentation)
- Beverages in open containers

Most testing centers require you to arrive 30-45 minutes before your scheduled exam time to complete check-in procedures. The proctor will verify your identity, explain test rules, and assign you a seat. Some facilities may require you to store personal items in a locker before entering the testing room.

Regarding breaks, policies vary by location. Many exams run without scheduled breaks, though you can typically use the restroom if needed. Your test clock will continue to run during these unscheduled breaks, so use them sparingly. Some testing centers may require you to sign out and in for breaks and may check your materials upon return.

The total exam duration is typically 4 hours. This may seem like plenty of time for 80 questions, but many test-takers find themselves watching the clock nervously toward the end. The combination of complex questions and code lookups can consume time quickly.

I remember checking my watch about halfway through my exam and realizing I'd only completed a third of the questions. That moment of panic could have derailed my performance, but I remembered Frank's advice: "The test is a marathon, not a sprint. Keep moving forward, and pick up speed where you can."

Arriving fully prepared for these logistics means you can focus your mental energy where it belongs—on answering the questions correctly, not worrying about rules or procedures. Knowing what to expect removes one more variable from an already stressful day.

1.3 Best Practices for Success

Nothing works if you don't. You can't skip studying and then expect to ace the exam just because you've done electrical work for years; even the best-prepared electricians can trip up on test day if they neglect some basic practices.

Arriving Early

I pulled into the test center parking lot 45 minutes before my scheduled exam. At first, I felt silly sitting in my truck with so much time to spare. But as I watched other test-takers rush in—some looking frantic as the start time approached—I knew I'd made the right call.

Arriving at least 30 minutes early gives you several advantages:

- Time to find the right building, room, and check-in location.
- A chance to use the restroom before the exam begins.
- A buffer for unexpected traffic or parking issues.
- A few minutes to flip through your code book and get mentally prepared.
- The opportunity to pick a good seat if seating is first-come, first-served.

I've known smart, skilled electricians who failed their exam because they arrived late and flustered. Your brain needs time to settle before tackling complex electrical questions. Give yourself that gift by planning your route to the test center, checking traffic reports, and setting multiple alarms if needed.

1.4 Handling Exam Stress

Test anxiety hits even the most experienced electricians. I know of a journeyman with 15 years of field experience who froze up when he opened his exam booklet. His hands shook so badly he could barely hold his pencil. Knowledge wasn't his problem—stress was.

Staying Calm Under Pressure

When your heart races and your mind goes blank, try these techniques that have helped many electricians get through the exam:

- **The 4-7-8 Breathing Method**: Take a deep breath for 4 seconds, hold it for 7 seconds, then exhale slowly for 8 seconds. This pattern activates your parasympathetic nervous system, which counters the fight-or-flight response. I used this technique several times during my exam, especially when facing questions that initially stumped me.
- **The "Name Five Things" Technique**: If your anxiety spikes during the exam, pause and name five things you can see, four things you can touch, three things you can hear, two things you can smell, and one thing you can taste. This focused grounding will pull your mind back to the present moment.
- **Positive Self-Talk**: Replace thoughts like "I'm going to fail" with "I've prepared for this" or "I know how to find the answers." Your internal dialogue shapes your performance more

than you might think.

- **The Paragraph Method**: If a question seems overwhelming, write a small paragraph explaining how you would approach it in the field. This often triggers your practical knowledge and leads you to the right answer.
- **Strategic Breaks**: If allowed, take a 30-second break every hour. Close your eyes, roll your shoulders, and reset your focus. This brief pause can prevent mental fatigue from building up.

Remember, some nervousness is normal and even helpful. A slight edge of adrenaline can sharpen your focus. The goal isn't to eliminate all stress but to keep it at a manageable level.

Time Management Strategies

You usually get three types of test-takers: those who race through and finish early (often missing details), those who get stuck on hard questions and run out of time, and those who pace themselves and complete the full exam. You want to be in that third group.

Here's a time management approach that works for many successful test-takers:

1. **The Three-Pass System**:

   * First pass (2 hours): Answer all questions you can solve quickly (under 2 minutes each). Skip anything that would take longer, but mark it for review.
   * Second pass (1 hour): Return to mark questions that need moderate effort.
   * Final pass (1 hour): Tackle the most difficult questions and review any you're uncertain about.

2. **The 3-Minute Rule**: If you can't make progress on a question after 3 minutes, mark it and move on. Return to it later with fresh eyes.

3. **Strategic Grouping**: Group similar questions mentally. If you just looked up a code article for one question, check if any nearby questions might use the same article before closing your book.

4. **Progress Tracking**: Note your pace at several points. For an 80-question test, aim to complete 20 questions per hour. If you fall behind, adjust by spending less time on each remaining question.

5. **Last-Call Strategy**: In the final 15 minutes, make sure every question has an answer marked, even if you must guess on a few. Remember, unanswered questions are guaranteed zero points.

The test isn't just checking if you know the material—it's checking if you can work under pressure while following the rules. Managing your time effectively is what demonstrates this critical skill. The exam tests both your electrical knowledge and your ability to perform under pressure. While these stress management practices might not guarantee a perfect score of 100% for the exam, they will help you show up ready not just to take the test but to master it.

2

National Electrical Code (NEC) Essentials

On my first day as a journeyman, I had to troubleshoot a complex control system at a manufacturing plant. The system had been acting up for weeks, and two other electricians had already taken a crack at it without success. The plant manager was hovering nearby, clearly doubtful that this new guy with a fresh license could solve what had stumped the veterans.

I opened my toolbox and pulled out my well-worn NEC handbook. The manager raised an eyebrow.

“You’re going to fix my machine by reading a book?” he asked, not hiding his skepticism.

I smiled and said, “Sometimes the answer isn’t in the wires—it’s in understanding why they’re arranged the way they are.”

Three hours later, we had the system running again. The issue wasn’t a failed component but an installation that didn’t meet code requirements for that specific application. The control wiring ran too close to the power conductors without proper separation and created an intermittent interference that baffled everyone.

The NEC is a troubleshooting guide, yes, but it’s also a design manual, and a safety handbook all rolled into one. The electricians who came before me knew how to install and repair equipment, but they hadn’t connected the dots back to the fundamental principles codified in those pages.

Throughout my career, I’ve seen the pattern repeat itself. The electricians who truly master the NEC don’t just memorize it—they understand the reasoning behind each requirement. They see the connections between different articles and know how to apply the code’s logic to situations that might not be explicitly covered.

In this chapter, we'll break down the most test-relevant sections of the NEC and discuss not just what the requirements are, but why they exist and how they work together. My goal isn't just to help you pass the test, but to develop the kind of deep understanding that will serve you throughout your career, just as it's served me.

2.1 Purpose and Organization of the NEC

The NEC is the backbone of electrical safety in the United States and sets the standard for safe electrical installations. Understanding the structure of the NEC will help you quickly find the information you need, whether you're on the job or studying for your exam.

Purpose of the NEC

The NEC exists for one primary reason—to keep people safe. Every article, table, and footnote ties back to this fundamental purpose. When I started in the trade, I viewed the code as a bunch of annoying rules that slowed down my work. It took a few close calls and one particularly patient inspector to help me see the bigger picture.

The main functions that it exists to serve are:

- **To ensure safe and reliable electrical installations.** Each requirement addresses a specific hazard that real electricians discovered through hard-won experience, often after someone got hurt or something caught fire. The three-foot working space in front of panels? That rule exists because electricians were getting trapped against live parts. The details about wire insulation types? Each one represents lessons learned about what fails under different conditions.
- **To protect people and property from electrical hazards**. The code doesn't just worry about electrical workers—it considers everyone who might come near your work. Homeowners who might stick a fork in an outlet, maintenance workers who need to access equipment years later, and firefighters who might spray water around electrical equipment during an emergency all benefit from your code compliance.
- **To standardize electrical practices across the country.** I once worked with a crew from Texas on a job in Pennsylvania. Despite the regional differences in how they spoke and what they called certain tools, we all understood what needed to happen because we shared the common language of the NEC. This standardization makes our work safer and more efficient across state lines.

I remember working on a renovation job where the building owner kept pushing us to skip some of the more expensive code requirements. "No inspector will ever see inside that wall," he told us. My foreman just pointed to a burn mark on his arm and said, "The code isn't about passing inspection—it's about making sure I go home with the same number of fingers I showed up with." That stuck with me. So, during your exam, remember this purpose because it can help you reason through questions when you're not sure of the exact code reference.

2.2 How the NEC Is Organized

When you open the NEC for the first time, it'll feel a little like trying to read a foreign language. The book is thick, the typography is tiny, and finding anything specific seems almost impossible. But think of the code like a map. The chapters are states, the articles are cities, and the sections are street addresses. Once you know the system, you can find anything. The NEC follows a logical organization that, once understood, makes navigation much simpler:

At the highest level, the NEC divides into nine chapters, with each chapter covering a distinct aspect of electrical work:

- **Chapter 1 (Articles 90-110)**: General rules that apply to all installations
- **Chapter 2 (Articles 200-285)**: Wiring and protection requirements
- **Chapter 3 (Articles 300-399)**: Wiring methods and materials
- **Chapter 4 (Articles 400-490)**: Equipment for general use
- **Chapter 5 (Articles 500-590)**: Special occupancies (hazardous locations, healthcare, etc.)
- **Chapter 6 (Articles 600-695)**: Special equipment (signs, elevators, etc.)
- **Chapter 7 (Articles 700-770)**: Special conditions (emergency systems, standby power, etc.)
- **Chapter 8 (Articles 800-840)**: Communications systems
- **Chapter 9**: Tables and examples

Within each chapter, individual articles tackle specific topics. For example, Article 430 covers motors, Article 250 handles grounding and bonding, and Article 310 addresses conductors for general wiring.

Each article then breaks down into sections using a decimal system. Take Article 210 on branch circuits as an example:

- 210.1 gives the scope of the article
- 210.3 provides rating requirements
- 210.8 covers GFCI protection
- 210.11 addresses required branch circuits

These sections often divide further into subsections using letters:

- 210.8(A) covers GFCI requirements for dwelling units
- 210.8(B) covers GFCI requirements for other than dwelling units

The organization gets even more detailed with numbered lists under many subsections. This hierarchical structure helps you drill down from general topics to very specific requirements.

I found that understanding this organization saved me countless hours during exam prep. Instead of flipping randomly through pages, I could make educated guesses about where to find information. Need motor overload protection requirements? That's likely in the 430s. Looking for junction box fill calculations? Check the 314s.

What helped me most was learning the chapter themes. If you know which chapter covers your topic, you've narrowed your search significantly. For example, knowing that Chapter 2 covers wiring and protection means questions about circuit breakers, grounding, and branch circuits will lead you there.

The code also includes helpful navigational tools:

- The table of contents at the front lists all articles.
- The index at the back helps find specific topics.
- Each article begins with a scope section that explains what it covers.
- Cross-references appear throughout to point you to related sections.

On my exam, I saved valuable minutes by knowing exactly where to look. For instance, when faced with a question about healthcare facility wiring, I didn't waste time searching random pages—I went straight to Article 517, knowing that special occupancies live in Chapter 5.

This organizational knowledge doesn't just help with the exam—it makes you more efficient on the job, too. When an inspector cites a code violation, knowing how to quickly find and understand the relevant section can save you from rework and help you explain your position if you believe you comply.

2.3 How to Use the NEC Effectively

The code book doesn't do you any good if you can't find what you need when you need it. This is something that I learned the hard way during an early job when the inspector asked me about a specific detail of a kitchen installation. I knew I'd read something about it, but spent 10 minutes flipping through pages while he watched, unimpressed.

Make the Code Book Your Own

We all have different learning styles, different ways of engaging with and interpreting learning content, so the most useful NEC is one that's personalized to you. When I got my first code book, Frank showed me his—dog-eared, highlighted, with tabs sticking out everywhere. "This isn't just a book," he said. "It's a tool that works better the more you use it."

First, mark the sections that you use most often:

- Article 210—Branch Circuits
- Article 250—Grounding and Bonding
- Article 300—Wiring Methods
- Article 310—Conductors
- Article 408—Switchboards and Panelboards
- Article 430—Motors

Use colored tabs to mark these pages for quick access. Color-coding helps—I use blue tabs for wiring methods, yellow for grounding, and red for overcurrent protection.

Highlighting should be selective and purposeful. Too much highlighting defeats the purpose. Mark key phrases, exceptions, and tables that you regularly reference. When I highlight, I focus on numbers (like minimum wire sizes or maximum distances) and exceptions to general rules.

Write notes in the margins, but keep them brief and clear. I note differences between the current code and previous versions that I've worked with, or reminders about local amendments that modify certain requirements.

Navigation Techniques

The NEC includes built-in tools to help you find information quickly:

The index at the back of the book should be your first stop when looking for an unfamiliar topic. If you need to find requirements for hot tubs, the index will direct you to Article 680. Learning to use the index effectively can cut your search time in half.

The table of contents provides a broad overview of where topics are located. Scanning the article titles can help you narrow down where to look.

The first few sections of each article usually explain its scope and provide general information. Reading these sections gives you context for the more detailed requirements that follow.

Cross-references appear throughout the text and point you to related information in other sections. These connections help you understand how different parts of the code work together.

I've found that understanding the code's organizational patterns speeds up lookups. For example, Section x.1 typically defines the scope, x.2 often covers definitions, and x.3 frequently addresses general requirements. Once you recognize these patterns, you can jump to the most likely section more quickly.

2.4 Key Code Sections Every Electrician Must Know

The NEC contains thousands of pages of rules and regulations, but certain sections are more important for electricians to be familiar with than others. Knowing these sections will help you stay safe and compliant while on the job and will ensure you're prepared for your exam.

Article 100-Definitions

The first time I failed to check a definition, it cost me a full day of rework. I had installed a "service disconnect" based on my understanding of what that meant, only to have the inspector point out that my installation didn't meet the NEC definition.

Article 100 serves as the NEC's dictionary. These aren't casual definitions—they're precise technical terms with specific meanings that might differ from everyday usage. Let me break down some of the most critical definitions you'll need to know:

Grounded vs. Grounding

These similar-sounding terms cause major confusion:

- **Grounded**: Refers to a conductor or component that is intentionally connected to the earth or to a conductive body that serves as earth. This is typically your neutral (white) conductor in most residential systems.
- **Grounding**: Refers to a conductor or component that connects equipment to the ground but doesn't normally carry current. This is your equipment grounding conductor (usually bare or green), which only carries current during a fault condition.

Why it matters: Installing a grounding conductor where a grounded conductor belongs can create dangerous conditions. For example, connecting a neutral load wire to ground at any point other than the main service creates parallel return paths and potential shock hazards.

2.5 Identified (for a specific purpose)

The NEC defines "identified" as recognizable as suitable for a specific purpose, function, use, or environment. This goes beyond simple marking or labeling.

For example, wire with white or gray insulation is "identified" as a grounded conductor. A terminal with a green screw is "identified" for connection to an equipment grounding conductor.

Why it matters: Using components that aren't identified for specific applications violates code, even if they might physically work. An unmarked box isn't identified for wet locations, even if it seems waterproof.

Service vs. Feeder

- **Service**: The conductors and equipment connecting the utility supply to the wiring system of the premises served. Service conductors extend from the utility connection point to the service disconnecting means.
- **Feeder**: All circuit conductors between the service equipment (or other power supply source) and the final branch-circuit overcurrent device.

Why it matters: Services have different requirements than feeders for overcurrent protection, disconnecting means, and grounding. I've seen contractors treat a subpanel feeder like a service, installing a second ground rod system (violating 250.32 in most cases).

Location Categories

- **Dry Location**: Not normally subject to dampness or wetness. A location classified as dry may be temporarily subject to dampness or wetness during construction.
- **Damp Location**: Partially protected locations subject to moderate degrees of moisture, such as under canopies, roofed open porches, and interior areas subject to condensation.
- **Wet Location**: Installations underground or in concrete slabs in direct contact with the earth; locations subject to saturation with water or other liquids; unprotected locations exposed to weather.

Why it matters: Equipment and wiring methods must match the location category. Using indoor-rated equipment in a damp location creates safety hazards and code violations. For example, "WR" receptacles are specifically identified for wet locations.

Article 110—Requirements for Electrical Installations

Article 110 covers the foundational rules that apply to almost all electrical work. Think of it as the general rules of the road before you get to specific traffic situations. This article appears frequently on the journeyman exam because it establishes core principles that affect nearly everything we do.

Working Space Requirements (110.26)

Section 110.26 specifies the clearances required around electrical equipment to allow safe operation and maintenance. The requirements are precise and non-negotiable:

- **Depth of Working Space (110.26(A)(1))**: The minimum clear distance from exposed live parts varies based on voltage:

  * 0-150V: 3 feet (condition 1), 3 feet (condition 2), 3 feet (condition 3)
  * 151-600V: 3 feet (condition 1), 3.5 feet (condition 2), 4 feet (condition 3)

The "conditions" refer to what's on the opposite wall:

- Condition 1: Exposed live parts on one side, no live or grounded parts on the other
- Condition 2: Exposed live parts on one side, grounded parts on the other
- Condition 3: Exposed live parts on both sides
- **Width of Working Space (110.26(A)(2))**: The width must be at least 30 inches or the width of the equipment, whichever is greater. This ensures you have room to work without contorting your body.
- **Height of Working Space (110.26(A)(3))**: The working space must extend from the floor to at least 6.5 feet or the height of the equipment, whichever is greater. No storage is permitted in this space.
- **Dedicated Equipment Space (110.26(E))**: For indoor installations, a dedicated space equal to the width and depth of the equipment, extending from the floor to 6 feet above the equipment, must be kept clear for electrical equipment.

Conductor Termination Provisions (110.14)

This section addresses how conductors connect to terminals and splices:

- **Temperature Limitations (110.14(C))**: This is one of the most misunderstood parts of the code. The ampacity of a conductor must be determined based on the lowest temperature rating of any connected termination, conductor, or device.

For most residential and commercial applications with circuits rated 100 amps or less, you must use the 60°C column of Table 310.16, even if you're using 75°C wire, unless all terminations are rated for 75°C. For circuits over 100 amps, you can use the 75°C column if all terminations are rated for 75°C.

- **Connector Types**: Pressure connectors and soldered joints must be made with identified materials and properly installed.
- **Tightening Torque (110.14(D))**: Terminals for more than one conductor must be identified for such use. Most terminals are designed for a single conductor, and doubling up creates loose connections and fire hazards.

I once had to rewire six panels because another electrician had doubled up neutrals on single terminal bars. The inspector caught it, citing 110.14.

Equipment Identification and Warning Signs (110.21)

- **Manufacturer's Markings (110.21(A))**: Equipment must be marked with the manufacturer's name, trademark, or other descriptive marking. This helps with traceability and ensuring proper application.
- **Field-Applied Warning Labels (110.21(B))**: If the equipment needs field-marked warning signs, they must be durable enough to withstand the environment. Paper labels in damp locations don't cut it.
- **Specific Markings**: Different equipment requires specific markings—service disconnects must be marked as such, and panelboards need to identify which circuits they control.

On a commercial job, we had to reinstall dozens of warning labels because the painter had removed them during finishing. The labels weren't just for appearance—they provided critical safety information.

Installation and Use Requirements (110.3)

- **Examination of Equipment (110.3(A))**: Equipment must be evaluated for specific characteristics, including suitability for installation, mechanical strength, wire-bending and connection space, and electrical insulation.
- **Installation and Use (110.3(B))**: This seemingly simple requirement is one of the most important in the entire code. It states that "listed or labeled equipment shall be installed and used under any instructions included in the listing or labeling."

Other Critical Parts of Article 110

- **Deteriorating Agents (110.11)**: Equipment must be suitable for the environment where it's installed. This includes protection from moisture, temperature extremes, and chemical exposure.

- **Mechanical Execution of Work (110.12)**: Electrical equipment must be installed in a "neat and workmanlike manner." While somewhat subjective, this gives inspectors latitude to fail sloppy work.
- **Interrupting Rating (110.9)**: Equipment intended to interrupt current must have an interrupting rating sufficient for the voltage and current available at the terminals. This prevents the installation of breakers that can't safely interrupt available fault current.
- **Circuit Impedance and Other Characteristics (110.10)**: Overcurrent protection, conductors, and connected components must be selected and coordinated to properly clear faults without extensive damage.

Article 110 seems basic at first glance, but its requirements are fundamental to electrical safety. The exam frequently includes questions about working spaces, conductor terminations, and the obligation to follow the manufacturer's instructions. Knowing these requirements will serve you well on both the test and the job site.

2.6 Article 250: Grounding and Bonding

Article 250 might arguably be the most important article in the entire code. Grounding and bonding mistakes kill people. I've seen more dangerous installations from misunderstood grounding requirements than from any other cause.

The Difference Between Grounding and Bonding (250.4)

The distinction between grounding and bonding is fundamental yet frequently confused:

- **Grounding** is connecting an electrical system or equipment to the earth. This provides a reference point for voltage and helps protect against lightning and static electricity. The grounded conductor (neutral) connects to ground at the service equipment, creating a zero-voltage reference point.
- **Bonding** is connecting conductive parts to establish electrical continuity and conductivity. Bonding ensures that connected parts share the same electrical potential, eliminating voltage differences that could create shock hazards.

Section 250.4 separates these requirements into two categories:

- **Grounded Systems (250.4(A))**: These are systems with a conductor (usually the neutral) intentionally grounded. The requirements include:

- Maintaining the electrical system ground connection (250.4(A)(1))
- Providing a low-impedance fault current path (250.4(A)(3))
- Bonding the normally non-current-carrying metal parts to the grounded conductor (250.4(A)(4))

- **Ungrounded Systems (250.4(B))**: These systems have no intentionally grounded conductor but still require:
  - Grounding of electrical equipment (250.4(B)(1))
  - Bonding to establish an effective fault current path (250.4(B)(3))
  - Limiting voltage to ground during normal operation (250.4(B)(4))

Sizing Grounding Electrode Conductors (250.66)

The grounding electrode conductor connects your grounding electrode system (ground rods, water pipes, etc.) to your service equipment. Its size depends on the size of the service conductors:

- **Table 250.66**: This table specifies minimum sizes based on the largest service-entrance conductor:
  - For copper service conductors 2 AWG or smaller: 8 AWG copper GEC
  - For copper service conductors 1/0 to 3/0: 6 AWG copper GEC
  - For copper service conductors over 3/0 through 350 kcmil: 4 AWG copper GEC
  - For copper service conductors over 350 kcmil through 600 kcmil: 2 AWG copper GEC
  - For copper service conductors over 600 kcmil through 1100 kcmil: 1/0 AWG copper GEC
  - For copper service conductors over 1100 kcmil: 2/0 AWG copper GEC
- **Maximum Size Requirements (250.66(A)-(C))**: For specific types of grounding electrodes, the GEC need not be larger than:
  - Ground rod or pipe electrode: 6 AWG copper (250.66(A))
  - Concrete-encased electrode: 4 AWG copper (250.66(B))
  - Ground ring: The size of the ground ring conductor (250.66(C))

These maximum size requirements reflect the fact that the grounding electrode's connection to earth is the limiting factor, not the conductor size. Even with a 2000-amp service, you only need a 6 AWG copper conductor to a ground rod, because the rod's earth contact is the limiting factor.

Equipment Grounding Conductor Sizing (250.122)

The equipment grounding conductor (EGC) provides a path for fault current to return to the source, allowing overcurrent devices to trip quickly. Its size is based on the rating of the overcurrent device:

- **Table 250.122**: This table specifies the minimum EGC size based on the circuit's overcurrent protection:
    * 15A circuit: 14 AWG copper EGC
    * 20A circuit: 12 AWG copper EGC
    * 60A circuit: 10 AWG copper EGC
    * 100A circuit: 8 AWG copper EGC
    * 200A circuit: 6 AWG copper EGC
    * 400A circuit: 3 AWG copper EGC
- **Increased in Size (250.122(B))**: If ungrounded conductors are increased in size for voltage drop or other reasons, the EGC must be proportionally increased in size.

For example, if you increase a 10 AWG circuit conductor to 8 AWG for voltage drop, the EGC must also be increased proportionally. Since 10 AWG has a cross-sectional area of 10,380 circular mils and 8 AWG has 16,510 circular mils, the ratio is 16,510 ÷ 10,380 = 1.59. Multiply the original EGC size in circular mils by this ratio to get the new minimum size.

This proportional increase requirement ensures the EGC can handle fault currents for the circuit's entire length, accounting for voltage drop in the fault path.

- **Multiple Circuits (250.122(C))**: A single EGC can serve multiple circuits in the same raceway or cable, but it must be sized for the largest overcurrent device.
- **Flexible Cords and Cables (250.122(E))**: For flexible cords with equipment grounding conductors, the EGC must comply with 250.122, except for listed equipment such as fixtures where the supply cord includes a smaller EGC.

2.7 Article 300: Wiring Methods

Article 300 addresses how we install wiring regardless of the specific method used. It's like the common rules that apply whether you're using conduit, cable, or open wiring. This article comes into play on nearly every job you'll ever work, making it essential knowledge for both your exam and career.

Protection Against Physical Damage (300.4)

Physical damage protection requirements are some of the most commonly cited violations I see in the field. These rules protect wiring from nails, screws, and other hazards:

- **Cables Through Wood Members (300.4(A))**: When running NM cable, AC cable, or MC cable through holes in wood framing members, you must keep the edge of the hole at least 1¼ inches from the nearest edge of the stud or joist. If you can't maintain this distance, you must protect the cable with a 1/16-inch steel plate or sleeve.

This requirement exists because drywall installers and finish carpenters often drive fasteners into the edges of framing members. That steel plate could be all that stands between a cable and a nail or screw.

- **Cables Through Metal Members (300.4(B))**: Cables running through metal framing members must be protected by bushings or grommets that cover all metal edges. Metal studs have sharp edges that can cut through insulation over time as cables expand and contract with heating and cooling cycles.
- **Cables in Parallel to Framing Members (300.4(D))**: Cables running parallel to framing members must be protected by steel plates if they're within 1¼ inches of the edge where nails or screws are likely to penetrate. Many electricians forget this requirement when running cables along the bottom of floor joists.
- **Cables in Shallow Grooves (300.4(E))**: Cables installed in shallow grooves must be protected by at least a 1/16-inch steel plate. I've seen many kitchen remodels where contractors tried to embed cables in shallow wall channels without proper protection.

Conductor Requirements (300.3)

These requirements address how conductors must be installed within raceways and enclosures:

- **Conductors of the Same Circuit (300.3(B))**: All conductors of the same circuit, including the neutral and equipment grounding conductor, must be installed in the same raceway, cable, trench, or cord. This prevents inductive heating and voltage issues.

On commercial jobs, I occasionally see inexperienced electricians run the neutral for a circuit through a different conduit than the phase conductors. This violates 300.3(B) and can create dangerous inductive heating in metal raceways.

- **Conductors of Different Systems (300.3(C))**: Conductors of AC and DC systems operating at 600 volts or less can occupy the same equipment wiring enclosure, raceway, or cable if all conductors have an insulation voltage rating at least equal to the maximum circuit voltage.

However, low-voltage control wiring, fire alarm, or communications cables require separation from power conductors unless adequate barriers are installed.

- **Individual Conductors (300.3(A))**: Single, insulated conductors must be installed in a raceway, cable, or enclosure. You can't just run individual THHN conductors through an open space.

Underground Installation Requirements (300.5)

Underground wiring has specific depth and protection requirements that vary based on the installation method:

- **Minimum Cover Requirements (Table 300.5)**: Direct-buried cables or conductors require a minimum cover of:
  * 24 inches in areas not subject to vehicle traffic
  * 18 inches for residential branch circuits rated 120V or less with GFCI protection and maximum overcurrent protection of 20A
  * 24 inches under driveways and parking areas of single-family dwellings
  * 18 inches under 2 inches of concrete
  * 12 inches under 4 inches of concrete

- **Protection from Damage (300.5(D))**: Direct-buried cables and conductors emerging from the ground must be protected by enclosures or raceways extending from the minimum cover distance below grade to at least 8 feet above grade. If run up a pole, the protection must extend at least 8 feet above grade.
- **Underground Splices (300.5(E))**: Direct-buried cables must use splicing methods identified for wet locations. Many electricians forget that underground is considered a wet location, even in dry climates.
- **Backfill (300.5(F))**: The first 6 inches of backfill must be free of rocks and sharp objects. I've seen cables damaged during backfilling when contractors used excavated soil containing rocks and debris.
- **Raceway Seals (300.5(G))**: Conduits or raceways entering a building from underground must be sealed to prevent moisture migration. This is often overlooked, but moisture can travel through raceways and cause equipment damage or create shock hazards.
- **Bushing (300.5(H))**: A bushing or termination fitting must protect direct-buried cables entering raceways. The sharp edge of a raceway can damage cable insulation over time.

Raceways Exposed to Different Temperatures (300.7)

- **Sealing (300.7(A))**: Where raceways pass from the interior to the exterior of a building or between areas of different temperatures, they must be sealed to prevent air circulation. This prevents condensation from forming inside the raceway when warm, humid air meets a cooler surface.

- **Expansion Fittings (300.7(B))**: Raceways must include expansion fittings where necessary to compensate for thermal expansion and contraction. Long runs of exposed conduit on rooftops need expansion fittings to prevent pulling apart at terminations or damaging connected equipment.

Other Important Provisions in Article 300

- **Spread of Fire or Products of Combustion (300.21)**: Electrical installations in hollow spaces, vertical shafts, and ventilation ducts must be installed so the spread of fire is not substantially increased. This means properly firestopping penetrations through fire-rated walls, floors, and ceilings.
- **Securing and Supporting (300.11)**: Raceways, cable assemblies, and boxes must be securely fastened in place. You cannot use other raceways, pipes, or ceiling support wires to support electrical equipment unless designed for that purpose.
- **Raceway or Cable to Open or Concealed Wiring (300.15)**: A box or fitting is required at each conductor splice, outlet, switch point, junction, or pull point for concealed knob-and-tube wiring. This ensures that all connections remain accessible.

Article 300 provides the foundation for proper wiring installation, regardless of the specific wiring method used. Mastering these requirements will help you pass your exam and, more importantly, keep your installations safe and code-compliant throughout your career.

2.8 Article 400: Flexible Cords and Cables

Article 400 covers when and how you can use flexible cords. This may seem like a minor topic, but it appears frequently on exams because it's so often misunderstood in the field. I've found that flexible cord violations are among the most common issues cited during inspections.

Uses Not Permitted for Flexible Cords (400.8)

Section 400.8 explicitly prohibits several common but dangerous practices. You cannot use flexible cords:

- **As a Substitute for Fixed Wiring (400.8(1))**: Flexible cords cannot replace the permanent wiring of a structure. This is probably the most common violation I see—extension cords powering equipment that should have permanent outlets installed instead.

- **Through Holes in Walls, Ceilings, or Floors (400.8(2))**: Flexible cords cannot pass through holes in walls, suspended ceilings, dropped ceilings, or floors. This rule prevents cord damage from sharp edges and helps maintain fire ratings between spaces.
- **Through Doorways, Windows, or Similar Openings (400.8(3))**: Running cords through doorways exposes them to physical damage from closing doors and creates tripping hazards. Window openings similarly expose cords to damage and weather.
- **Attached to Building Surfaces (400.8(4))**: Cords cannot be attached to building surfaces. They cannot be stapled, nailed, or otherwise affixed to walls, ceilings, or structural members. I often find cords stapled along baseboards in older homes.
- **Concealed Behind Building Walls, Ceilings, or Floors (400.8(5))**: This prohibition prevents hidden damage from going undetected and helps prevent fire spread. Hidden cords can be damaged by nails, screws, or pests without anyone knowing until a fire starts.
- **Installed in Raceways (400.8(6))**: Flexible cords cannot be installed in raceways unless specifically permitted elsewhere in the code. Raceways trap heat and prevent proper cord cooling.
- **Where Subject to Physical Damage (400.8(7))**: Cords cannot be used where they would be exposed to physical damage. This includes high-traffic areas, places where heavy items might be placed on them, or locations with sharp edges.
- **Exceeding Voltage Limitations (400.8(8))**: Cords have specific voltage limitations based on their construction. Using a cord beyond its voltage rating creates shock and fire hazards.

These prohibitions exist because cord installations have caused numerous fires and electrical accidents. The flexibility that makes cords useful also makes them vulnerable to damage from abrasion, crushing, and environmental factors.

Using the Tables to Determine Cord Ampacity (400.5)

Determining the correct ampacity for flexible cords requires understanding several factors:

- **Table 400.5(A)(1)**: This table provides the ampacity of flexible cords based on conductor size and temperature rating. Three columns cover different temperature ratings: 60°C, 75°C, and 90°C.
- **Number of Current-Carrying Conductors (400.5(A)(3))**: When a cord contains more

than three current-carrying conductors, you must apply ampacity adjustment factors:

* 4-6 conductors: reduce ampacity to 80%
* 7-9 conductors: reduce ampacity to 70%
* 10-20 conductors: reduce ampacity to 50%
* 21-30 conductors: reduce ampacity to 45%
* 31-40 conductors: reduce ampacity to 40%

- **Temperature Derating (400.5(A)(2))**: For cords used in ambient temperatures above 30°C (86°F), you must apply temperature correction factors from Table 310.15(B)(2)(a).

Types of Cords and Their Applications (Table 400.4)

Table 400.4 lists all recognized cord types and their permitted uses. This table is crucial for selecting the right cord for specific applications.

Some common cord types include:

- **Type S, SO, ST**: These are hard service cords used for extra-hard usage. SO has an oil-resistant outer covering, while ST has a thermoplastic outer covering. They're suitable for most heavy-duty applications in dry and damp locations.
- **Type SJ, SJO, SJT**: These are junior hard service cords, lighter duty than S types but still suitable for many applications. They're often used for appliances and power tools.
- **Type SPT**: These are parallel, thermoplastic-jacketed cords often used for lamps and small appliances. They come in different thicknesses (SPT-1, SPT-2, SPT-3).
- **Type SEO**: This is elevator cable with an oil-resistant outer covering. It's specifically designed for elevator applications and has special requirements.

Each cord type has specific approvals for:

- Wet or damp locations
- Hard or extra-hard usage
- Sunlight resistance
- Oil resistance
- Temperature ratings
- Voltage limitations

For the exam, focus on knowing which cords are approved for:

- Outdoor use (those marked "WA" or "W")
- Oil-resistant applications (those with "O" in the designation)
- Hard usage vs. extra-hard usage

Other Important Provisions in Article 400

- **Flexible Cord and Cable Markings (400.6)**: Cords must be marked with the manufacturer's name, wire size, number of conductors, voltage rating, and type designation. These markings help ensure the cord is used within its design parameters.
- **Grounding Conductor Size (400.23)**: The size of the grounding conductor in flexible cords must comply with 250.122. This ensures adequate fault current capacity.
- **Bending Radius (400.10)**: Cords must not be bent at a radius less than specified by the manufacturer. Tight bends can damage insulation and cause conductor failure.
- **Pull at Joints and Terminals (400.10)**: Cords must be installed so that tension is not transmitted to the terminal connections. Strain relief must be provided at connections to prevent conductor damage.

2.9 Article 450: Transformers

Transformers connect to almost everything in our electrical system, making Article 450 critical knowledge. In my experience, many electricians handle transformer installations incorrectly because they don't fully understand these requirements.

Overcurrent Protection (450.3)

The requirements for transformer overcurrent protection vary depending on the transformer type, impedance, and primary/secondary voltage. This is one of the most tested areas on the journeyman exam.

- **Table 450.3(A)**: This table applies to transformers over 1000V and specifies the maximum setting for primary protection as a percentage of transformer-rated current.
- **Table 450.3(B)**: This table applies to transformers 1000V or less and is what you'll encounter most often. The requirements vary based on whether the transformer exceeds 9 amperes:

For transformers with primary current less than 9 amperes:

- Primary protection must be no more than 167% of the primary current rating
- Secondary protection must be no more than 167% of the secondary current rating
- If both primary and secondary protection are provided, each can be rated up to 250%

For transformers with primary current 9 amperes or more:

- Primary protection must be no more than 125% of the primary current rating
- Secondary protection must be no more than 125% of the secondary current rating
- If both primary and secondary protection are provided, the primary protection can be rated up to 250%

Transformer Vaults (450.41-450.48)

When transformers exceed certain ratings or contain flammable liquids, they must be installed in transformer vaults that meet stringent construction requirements:

- **Construction Requirements (450.42)**: Vaults must be constructed of fire-resistant materials with a minimum fire rating of 3 hours. The walls, roof, and floor must all meet this rating.
- **Doorways (450.43)**: Vault doors must swing out and be equipped with locks. They must be fire-rated for at least 3 hours and designed to contain fire and smoke.
- **Ventilation (450.45)**: Vaults must have adequate ventilation to remove heat generated by transformer losses. Openings must be covered with durable gratings, screens, or louvers.
- **Water Pipes and Accessories (450.47)**: Foreign systems, such as water pipes or HVAC ducts, are not permitted to enter or pass through a transformer vault unless necessary for vault equipment.
- **Storage in Vaults (450.48)**: Transformer vaults cannot be used for storage. I've seen maintenance crews using transformer vaults as storage closets, creating serious fire hazards.

Transformer Installation Requirements (450.8)

These requirements address the physical installation of transformers:

- **Wall, Column, or Floor Mounting (450.8(C))**: Dry-type transformers of 600 volts or less can be attached directly to walls, columns, or floors if they don't exceed 1000 VA. Larger units must be installed with a minimum 6-inch airspace between the transformer and combustible materials.
- **Attachment to Building Surfaces (450.8(A))**: Transformers mounted on building surfaces must be securely attached and installed to minimize transmission of vibration and noise.
- **Suspended Units (450.8(B))**: Transformers can be suspended from building structures if the mounting method is designed for the weight and has a sufficient safety factor.
- **Raceway Connections (450.8(E))**: Raceway connections to transformers must be flexible when the transformer is mounted on vibration isolators. This prevents vibration transmission to the raceway system and connected equipment.

Ventilation Requirements (450.9)

Transformers generate heat during operation, and proper ventilation is essential for safety and longevity:

- **Temperature Rise**: Transformers must be installed so that the temperature rise from heat under full load conditions will not damage the insulation system. Each transformer has a temperature rise rating that must not be exceeded.
- **Airflow Clearances**: Ventilation openings must provide sufficient airflow to dissipate heat. Obstructing these openings can cause overheating and premature failure.
- **Location**: Transformers should be installed in well-ventilated areas with sufficient space for air circulation on all sides.

As you prepare for your exam, practice moving through the code book quickly. Time yourself looking up specific requirements. Create scenarios that require you to find and apply multiple code sections. This practice will build both speed and confidence for test day.

3

Electrical Fundamentals

"This motor's toast," muttered the maintenance supervisor, wiping sweat from his forehead. "Third one this month. Must be getting junk from the supplier."

I'd been called in to troubleshoot a manufacturing line where 5HP motors kept burning out within weeks of installation. The plant was losing thousands in downtime, and tempers were running hot.

"Mind if I check something first?" I asked, pulling out my multimeter.

Ten minutes later, I showed him the reading: 452 volts on a 480-volt system. "Your voltage is low, but that's not what's killing your motors," I explained. "You've got a current imbalance of nearly 20% between phases. Those motors are running hot on one winding until they burn up."

The supervisor stared at me like I'd performed magic. "How'd you know to check that?"

"Because motors don't just die for no reason," I told him. "And phase imbalance is invisible unless you know to look for it."

That afternoon saved his company the cost of replacing another motor and hours of production downtime. Not because I knew some industry secret, but because I understood how three-phase power actually works.

The fundamentals of electricity aren't just textbook theory—they're practical tools that solve real problems and save real money. I've watched journeymen with years of experience waste hours replacing perfectly good parts because they didn't understand what was actually happening inside the wires.

The journeyman exam tests these fundamentals because they're the difference between an electrician who can follow diagrams and one who can think through problems. You'll face questions about power calculations, circuit analysis, and motor theory not to torture you with math, but because these concepts matter on actual job sites.

In this chapter, we'll cover the electrical principles that drive everything we do, from residential wiring to industrial controls. We'll focus on the concepts that appear most frequently on the exam and in the field, breaking them down in straightforward terms with examples from real-world situations I've encountered.

Because at the end of the day, understanding electrical fundamentals isn't about passing a test—it's about becoming the electrician who finds solutions when everyone else is stumped.

3.1 Core Principles of Electricity

Electricity powers our world, and understanding how it works is essential for any electrician. From current and voltage to resistance and power, these core concepts form the foundation of electrical work.

What is Electricity?

Electricity is the flow of electrons through a conductor. It's driven by the difference in voltage between two points.

When a water tank is positioned above ground level, the water inside has potential energy because of its height—similar to voltage in an electrical system. When you open a valve at the bottom, water flows through the pipe—like current flowing through a wire. If the pipe is narrow or has obstacles, it resists the water flow—just like electrical resistance limits current.

Electricity is measured in terms of three fundamental quantities:

- **Voltage (V)**: The electrical pressure that pushes electrons through a conductor, measured in volts (V). Think of it as the force driving the current. Higher voltage means more pushing force on the electrons.
- **Current (I)**: The actual flow of electrons through a conductor, measured in amperes (A) or amps. This is what does the work in an electrical circuit—lights the lamps, runs the motors, and heats the elements.
- **Resistance (R)**: The opposition to current flow in a circuit, measured in ohms (Ω). Every material has some resistance to current flow. Conductors like copper have low resistance, while insulators like rubber have extremely high resistance.

These three quantities work together according to Ohm's Law, which states that current equals voltage divided by resistance (I = V/R). This simple relationship explains most of what happens in the circuits we work with daily.

Most people think electricity moves at the speed of light, but that's not quite right. The electromagnetic field that carries electrical energy moves close to light speed, but the actual electrons in the wire move surprisingly slowly—sometimes just inches per minute. It's like a row of billiard balls—when you hit the first one, the last one moves almost instantly, but the balls themselves don't travel the length of the table.

3.2 Ohm's Law and Its Applications

Ohm's Law is one of the most essential formulas in electricity, and it helps electricians solve many practical problems. This law relates voltage, current, and resistance in a straightforward equation.

Understanding Ohm's Law

- **Ohm's Law**: V= I x R

This equation tells us that voltage (V) is equal to the current (I) multiplied by the resistance (R) in a circuit. While simple, this relationship is the foundation for analyzing electrical circuits.

From this basic formula, we can derive two other useful equations:

- Current: I = V/R
- Resistance: R= V/I

These formulas work with any units as long as you're consistent:

- Voltage (V) in volts
- Current (I) in amperes or amps
- Resistance (R) in ohms

For example, if you have a 120-volt circuit with 10 ohms of resistance, the current would be: $I = V \div R = 120V \div 10\Omega = 12A$

Ohm's Law applies to both AC and DC circuits, though with AC circuits we must consider additional factors like reactance and impedance for complete accuracy.

The relationship between voltage, current, and resistance remains constant in properly functioning circuits. When troubleshooting, unexpected values often indicate a problem with connections, components, or insulation.

For complex circuits with multiple paths, Ohm's Law applies to each individual path as well as to the overall circuit. This helps when analyzing parallel and series configurations.

3.3 Kirchhoff's Law and Its Applications

Kirchhoff's Current Law (KCL): The total current entering a junction must equal the total current leaving the junction. This law helps with circuit analysis, particularly when dealing with multiple paths for current flow.

In practical terms, this means that current doesn't disappear or accumulate at connection points. If 15 amps flow into a junction where three branch circuits connect, then the sum of the current in those three branches must also equal 15 amps.

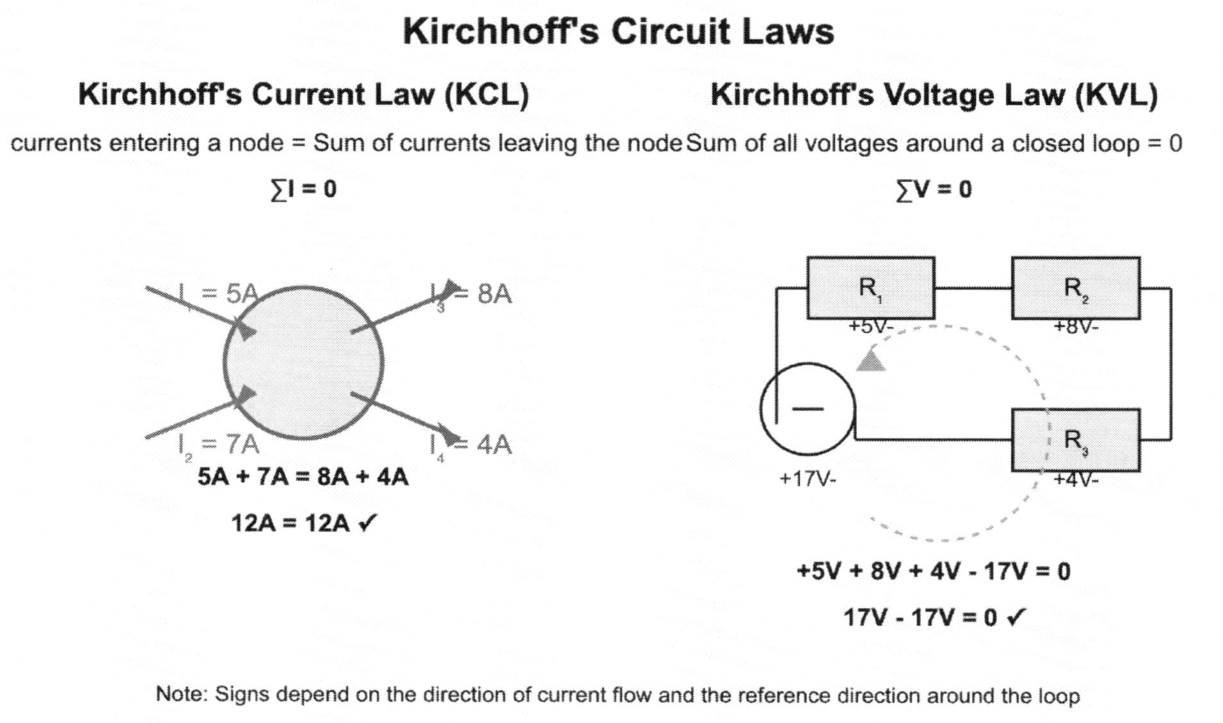

Figure 1: Kirchhoff's Circuit Laws

For example, in a panel with a 100-amp main feed that supplies four branch circuits drawing 20A, 30A, 15A, and 35A respectively, KCL confirms that 20A + 30A + 15A + 35A = 100A.

- Kirchhoff's Voltage Law (KVL): The sum of the voltages around any closed loop in a circuit must equal zero. This is because voltage is conserved in a closed loop.

This law explains why the voltage drops across series components add up to the source voltage. In a circuit with a 24V source and three loads in series, if two loads have voltage drops of 6V and 10V, the third load must have a voltage drop of 8V because 6V + 10V + 8V = 24V.

3.3 Common Electrical Formulas and Their Applications

When troubleshooting a failing 3-phase motor, calculating the capacity for a new service panel, or determining voltage drop on a long feeder run, you'll need more than just Ohm's Law in your toolkit. The formulas we're about to cover aren't just exam material—they're the calculations I use almost weekly to solve real problems in the field. These equations help you predict how electrical systems will behave before you install them, potentially saving thousands in materials and preventing dangerous overloads or undersized equipment.

Power Formula: $P = V \times I$

The power formula calculates how much electrical energy a device converts into other forms of energy (like heat, light, or mechanical motion) per second, measured in watts (W).

For DC circuits and AC circuits with unity power factor, the basic formula is: $P = V \times I$

Example 1:

When sizing a new circuit for a workshop, you need to know how much power your equipment will draw. If you're installing a 120V circuit for a table saw that draws 15A, you can calculate the power requirement: $P = V \times I = 120V \times 15A = 1{,}800W$ or 1.8kW

This tells you that the saw will consume 1,800 watts when running. This information helps you determine if your existing panel can handle the additional load and ensures you choose the right circuit breaker and wire size.

For AC circuits with a power factor less than 1, the formula becomes: $P = V \times I \times PF$

Example 2:

When working with a 240V single-phase air compressor motor that draws 12A and has a power factor of 0.85 (typical for many motors), you need to calculate the actual power consumption:

$P = V \times I \times PF = 240V \times 12A \times 0.85 = 2{,}448W$ or 2.45kW

This calculation is important because the utility charges for actual power (watts), not apparent power (volt-amps). Understanding the true power consumption helps estimate operating costs and properly size generators or inverters that might power this equipment.

For 3-phase circuits, the formula is: P = V × I × 1.732 × PF (where 1.732 is √3)

Electrical Energy Formula: E = P × t

The electrical energy formula helps calculate how much energy a device consumes over time, typically measured in kilowatt-hours (kWh). This is what appears on electric bills and determines operating costs.

E = P × t

Example 1:

A customer asks you to estimate the operating cost of a new 5kW electric water heater. You know it will run approximately 3 hours per day. To calculate the daily energy consumption: E = P × t = 5kW × 3h = 15kWh

Now you can estimate the monthly cost. If electricity costs $0.12 per kWh: Monthly cost = 15kWh × 30 days × $0.12/kWh = $54 per month

This helps the customer make an informed decision about their water heating options and possibly consider energy-saving alternatives.

Resistors in Parallel Formula: 1/R_total = 1/R_1 + 1/R_2 + ...

In parallel circuits, current has multiple paths to flow, reducing the overall resistance. The total resistance is always less than the smallest individual resistance.

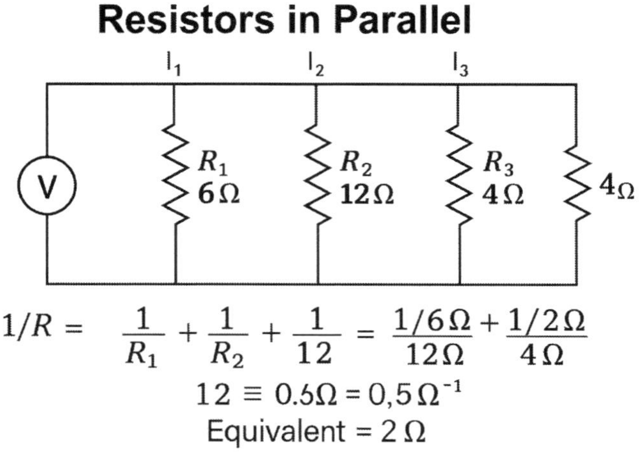

Figure 2: Resistors in Parallel

1/R_total = 1/R_1 + 1/R_2 + 1/R_3 + ...

Example 1:

When troubleshooting a heating system with two heating elements connected in parallel, you measure their individual resistances as 30Ω and 60Ω. To determine the total resistance:

$1/R_{total} = 1/30\Omega + 1/60\Omega = 2/60\Omega + 1/60\Omega = 3/60\Omega$ $R_{total} = 60\Omega \div 3 = 20\Omega$

With a 240V supply, you can now calculate the total current: $I = V \div R = 240V \div 20\Omega = 12A$

This helps you verify if the circuit breaker size is appropriate for this application and check if the system is operating correctly.

For just two resistors, you can use the shortcut formula: $R_{total} = (R_1 \times R_2) \div (R_1 + R_2)$ $R_{total} = (30\Omega \times 60\Omega) \div (30\Omega + 60\Omega) = 1{,}800\Omega \div 90\Omega = 20\Omega$

Resistors in Series Formula: R_total = R_1 + R_2 + ...

In series circuits, resistances add directly because current has only one path to follow. The total resistance equals the sum of all individual resistances.

$R_{total} = R_1 + R_2 + R_3 + \ldots$

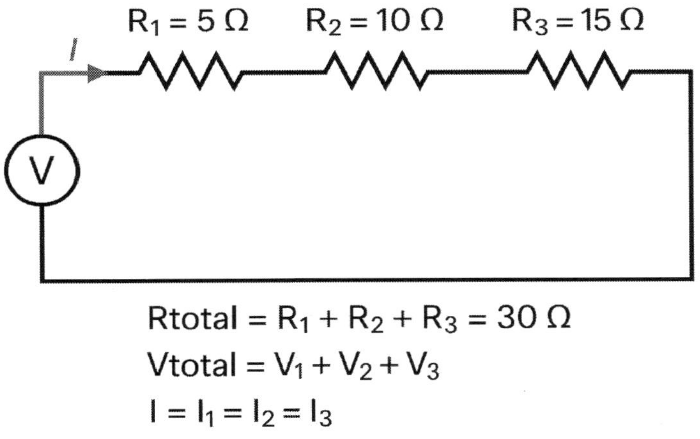

Figure 3: Resistors in series

Example 1:

When designing a custom control circuit, you need three specific resistance values in series to create a voltage divider. If the resistors have values of 5Ω, 8Ω, and 12Ω, the total resistance is:

$R_{total} = 5\Omega + 8\Omega + 12\Omega = 25\Omega$

With a 24V DC supply, the circuit will draw: $I = V \div R = 24V \div 25\Omega = 0.96A$

This calculation helps you select components with the appropriate power ratings and ensures the circuit functions as designed.

Example 2:

When installing a motor 150 feet from the panel, you need to calculate the total circuit resistance to check for potential voltage drop issues. Using 10 AWG copper wire (with a resistance of 0.1Ω per 100 feet) and knowing the motor has an impedance of 1.5Ω: R_total =

$(0.1\Omega \times 150ft \div 100ft) + 1.5\Omega = 0.15\Omega + 1.5\Omega = 1.65\Omega$

With this information, you can calculate the voltage drop: $V_{drop} = I \times R = 20A \times 0.15\Omega = 3V$

This 3V drop might be significant depending on the application. For a 240V circuit, this represents a 1.25% voltage drop, which is within the recommended limits. However, for a critical motor application, you might consider using larger conductors to minimize this drop.

3.4 Quiz: Testing Your Knowledge of Electrical Fundamentals

Now that you've learned the core principles of electricity, Ohm's Law, Kirchhoff's Laws, and common electrical formulas, it's time to test your knowledge. This quiz will help you assess your understanding of the material and identify areas that may need further review.

Quiz Overview

- Multiple-choice, true/false, and calculation-based questions to test your understanding.
- Topics will cover all major formulas and concepts discussed in the chapter.
- Detailed explanations will be provided for each question to reinforce learning.

Practice Questions

1. **A 120V circuit has a 15A load. What is the power consumption in watts?**

    a) 1,800W
    b) 135W
    c) 8W
    d) 105W

2. **What is the total resistance of three 30Ω resistors connected in series?**

   a) 10Ω
   b) 30Ω
   c) 60Ω
   d) 90Ω

3. **What is the total resistance of three 30Ω resistors connected in parallel?**

   a) 10Ω
   b) 30Ω
   c) 60Ω
   d) 90Ω

4. **A three-phase, 480V motor draws 25A with a power factor of 0.85. What is the power consumption?**

   a) 12,000W
   b) 17,688W
   c) 20,198W
   d) 25,500W

5. **According to Kirchhoff's Current Law, at any junction in a circuit:**

   a) The sum of all currents must equal zero
   b) Current must be the same throughout the circuit
   c) Voltage must be constant
   d) Resistance decreases with each branch

6. **A 3kW baseboard heater operates for 5 hours. How much energy does it consume?**

   a) 15kWh
   b) 0.6kWh
   c) 8kWh
   d) 3kWh

7. **True or False: In a parallel circuit, the total resistance is always greater than the smallest individual resistance.**

   a) True
   b) False

8. **If a 240V circuit has a total resistance of 20Ω, what is the current flow?**
   a) 4,800A
   b) 12A
   c) 260A
   d) 220A

9. **According to Kirchhoff's Voltage Law, in any closed loop of a circuit:**
   a) The sum of all voltage drops equals the supply voltage
   b) Current remains constant
   c) Resistance increases with distance
   d) Power is conserved

10. **Two branch circuits with impedances of 40Ω and 60Ω are connected in parallel. What is the total impedance?**
   a) 100Ω
   b) 50Ω
   c) 24Ω
   d) 2,400Ω

Answers and Explanations

1. **Answer: a) 1,800W** Using the power formula P = V × I = 120V × 15A = 1,800W.
2. **Answer: d) 90Ω** In series circuits, resistances add directly: R_total = 30Ω + 30Ω + 30Ω = 90Ω.
3. **Answer: a) 10Ω** For parallel resistors of equal value, divide the resistance by the number of resistors: R_total = 30Ω ÷ 3 = 10Ω.
4. **Answer: c) 20,198W** Using the 3-phase power formula: P = V × I × 1.732 × PF = 480V × 25A × 1.732 × 0.85 = 20,198W or about 20.2kW.
5. **Answer: a) The sum of all currents must equal zero** Kirchhoff's Current Law states that the algebraic sum of currents entering and leaving a junction must equal zero, or simply put, current in equals current out.
6. **Answer: a) 15kWh** Using the energy formula: E = P × t = 3kW × 5h = 15kWh.
7. **Answer: b) False** In a parallel circuit, the total resistance is always LESS than the smallest individual resistance, not greater.
8. **Answer: b) 12A** Using Ohm's Law: I = V ÷ R = 240V ÷ 20Ω = 12A.

9. **Answer: a) The sum of all voltage drops equals the supply voltage** Kirchhoff's Voltage Law states that the sum of all voltage drops around a closed loop equals the sum of all voltage sources in that loop.

10. **Answer: c) 24Ω** Using the parallel resistance formula: 1/R_total = 1/40Ω + 1/60Ω = 3/120Ω + 2/120Ω = 5/120Ω R_total = 120Ω ÷ 5 = 24Ω Alternatively, using the shortcut formula for two resistors: R_total = (R_1 × R_2) ÷ (R_1 + R_2) = (40Ω × 60Ω) ÷ (40Ω + 60Ω) = 2,400Ω ÷ 100Ω = 24Ω

Assessment

Count your correct answers:

- 9-10 correct: Excellent! You have a solid understanding of electrical fundamentals.
- 7-8 correct: Good job! Review the questions you missed.
- 5-6 correct: You're on the right track, but should review the chapter.
- Below 5: Take time to re-read the chapter and work through the examples again.

Remember that these concepts will appear throughout the exam, so it's worth ensuring you have a solid grasp of them before moving on.

4

Power Distribution and Sources

The manufacturing plant had been down for six hours when I arrived. A team of maintenance electricians stood around a dead distribution panel, scratching their heads. The plant manager paced nervously nearby, occasionally glancing at the idle production line costing them $10,000 per hour in lost revenue.

"We've checked every breaker twice," one of the facility's electricians told me. "The utility says it's not their problem. We're running out of ideas."

Starting at the service entrance, I methodically traced the power distribution system. The 480V, 3-phase main service showed balanced voltages on all phases. The main switchboard looked normal too. But at the subpanel feeding the affected area, I found something telling—two phases measured 277V to ground, but the third showed only 175V.

Working backward through the system, I eventually discovered a loose connection at a busway splice plate. The connection had overheated, partially melting the insulation and creating high resistance on one phase. It wasn't visible without thermal imaging, but the voltage readings exposed the problem.

Within thirty minutes, we had the connection cleaned, tightened, and properly torqued. The equipment hummed back to life, and production resumed.

This experience highlights an essential truth: power distribution isn't just about moving electricity from point A to point B. It's about understanding the entire system—how it's generated, delivered, transformed, and utilized. When you grasp these concepts at a fundamental level, troubleshooting becomes less about guesswork and more about logical deduction.

In this chapter, we'll explore how electricity gets from the utility to the outlets in a building. We'll examine different service types, grounding systems, and how to calculate loads. Whether you're working on residential services or industrial power systems, these principles apply universally.

Because when a customer is losing thousands of dollars per hour, the difference between a good electrician and a great one isn't just technical knowledge, it lies in their ability to see the entire power distribution system as a cohesive whole.

4.1 Types of Power Generation and Distribution

From massive nuclear plants generating gigawatts of power to solar panels on residential rooftops, electricity generation comes in many forms before reaching your service panel. As an electrician, you're the final link in a remarkable chain that might stretch hundreds of miles from source to outlet. While you'll rarely work directly on generation equipment, understanding how power is produced and delivered provides crucial context for your daily work—especially when dealing with backup systems, alternative energy sources, or power quality issues.

The path electricity takes before reaching your multimeter involves multiple voltage transformations, sophisticated switching systems, and an intricate web of transmission and distribution networks. Let's break down this journey from generation to utilization, focusing on what matters most for your work as a journeyman electrician.

Conventional Power Generation

- **Fossil Fuels**: Power plants burn coal, oil, or natural gas to generate steam, which drives turbines connected to generators. These plants provide about 60% of U.S. electricity generation and can produce consistent power regardless of weather conditions. The generated voltage typically starts at around 13.8kV to 24kV at the generator terminals.
- **Nuclear Power**: Nuclear reactors use controlled nuclear reactions to produce heat, which is then converted to electricity. The heat from nuclear fission creates steam to drive turbines, similar to fossil fuel plants. Nuclear plants generate about 20% of U.S. electricity and produce minimal air pollution or greenhouse gases during operation. These facilities typically generate power at 22kV to 24kV before stepping up to transmission voltages.
- **Hydroelectric Power**: Water flow is harnessed through dams to drive turbines that generate electricity. Hydroelectric facilities can quickly adjust output to match demand and serve as reliable baseline power sources where water supply is consistent. They generate approximately 7% of U.S. electricity. The generators in hydroelectric plants typically produce voltages between 13.8kV and 24kV, depending on the facility's size.

Each of these generation methods creates alternating current (AC) electricity, which is then increased to extremely high voltages (138kV to 765kV) for long-distance transmission. The high voltage reduces line losses during transmission by decreasing the current for the same amount of power.

As an electrician, you'll rarely work directly with generation equipment, but understanding how electricity is produced helps you grasp the entire system from source to outlet. This knowledge becomes particularly valuable when working on emergency power systems or troubleshooting complex distribution issues.

4.2 Renewable Power Generation

The landscape of power generation has significantly improved in recent years, with renewable sources becoming more mainstream. These systems create unique challenges and opportunities as well for if you need to understand their distinct characteristics and integration requirements.

Solar Power

Solar panels capture sunlight and convert it directly into electricity using photovoltaic cells. These systems generate DC power that must be converted to AC through inverters before connecting to the grid or building systems.

Working with solar installations requires specialized knowledge beyond standard electrical work:

- PV arrays typically generate DC power at 600-1000V in commercial applications
- String inverters, microinverters, or power optimizers convert DC to grid-compatible AC
- NEC Article 690 governs solar PV installations with specific requirements for disconnects, grounding, and labeling
- Rapid shutdown systems provide emergency power cutoff to protect first responders

Modern solar systems often include monitoring capabilities that allow owners to track production and system health. As an electrician, you should understand how these components integrate with building electrical systems and utility interconnections.

Wind Power

Wind turbines convert the kinetic energy of wind into mechanical energy, which is then turned into electricity. Unlike the steady output of conventional plants, wind generation varies with wind speed and direction.

Wind systems typically include:

- Generator assemblies that produce variable frequency AC based on wind speed.
- Power conversion equipment to create grid-compatible output.
- Sophisticated control systems for blade pitch, direction, and emergency braking.
- Step-up transformers to increase voltage for transmission (for larger installations)

Residential wind systems usually connect at 240V single-phase, while commercial wind farms may generate at medium voltages (12.47kV to 34.5kV) before connecting to transmission networks.

Geothermal Power

Heat from the Earth's core is used to generate steam to drive turbines in geothermal plants. These systems provide exceptionally reliable baseload power with minimal environmental impact where geological conditions permit.

Geothermal systems:

- Maintain consistent output regardless of weather conditions
- Operate with high availability factors (often 95%+)
- Typically generate at medium voltages similar to conventional steam plants
- Require specialized equipment to handle mineral-rich steam and water

While most electricians won't work directly with geothermal generation, understanding these systems becomes important when servicing facilities powered by geothermal sources or working in regions where geothermal is prevalent.

The growing adoption of renewable energy sources has created new specializations within the electrical trade. Most journeyman electricians now focus specifically on solar installation, wind farm maintenance, or integration of renewable systems with conventional infrastructure. As these technologies continue to expand, familiarity with renewable generation systems will be valuable throughout your career.

4.3 Power Distribution Systems

Ever consider what happens to electricity between a power plant and your service panel? The journey spans hundreds of miles, multiple voltage changes, and some of the largest infrastructure in the country. We operate at the final stage of this system, but understanding the entire distribution network makes us more effective troubleshooters and problem-solvers.

Essentially, the power grid is a transportation system: transmission lines are the highways, distribution networks are the local roads, and transformers are the interchanges between them. This massive system works continuously to deliver precisely calibrated power to your customers' facilities. When voltage fluctuations, momentary outages, or power quality issues arise, the cause often lies somewhere in this distribution chain.

Transmission Lines

High-voltage transmission lines carry electricity long distances from power plants to substations. These lines operate at extremely high voltages to minimize losses over distance.

Key characteristics of transmission systems include:

- Voltages typically range from 138kV to 765kV.
- Three-phase configuration with large conductors and substantial clearances.
- Distinctive lattice towers or monopoles supporting conductors via massive insulator strings.
- Right-of-ways maintained clear of vegetation and structures.

The transmission system is the backbone of the electrical grid and enables power to move between regions and balance loads across generating sources. While most electricians won't work directly on these systems, understanding transmission fundamentals helps explain why certain disturbances (like lightning strikes or equipment failures) can affect wide areas simultaneously.

Distribution Networks

After stepping down the voltage at substations, the electricity is distributed to homes and businesses through lower-voltage lines. This distribution network is where most electricians interface with the utility system.

The distribution system consists of:

- **Primary Distribution**: Medium voltage lines (typically 4kV to 35kV) delivering power to local areas
- **Distribution Transformers**: Step-down transformers converting medium voltage to utilization voltages
- **Secondary Distribution**: Lower voltage conductors (120/240V, 208Y/120V, or 480Y/277V) connecting to service entrances

Distribution systems may be overhead (on utility poles) or underground. Underground systems offer better aesthetics and weather resistance but cost more to install and repair. In commercial areas, pad-mounted transformers and switchgear are common, while residential areas typically use pole-mounted or smaller pad-mounted transformers.

The point where the distribution system connects to customer wiring (the service point) marks the transition from utility responsibility to customer/electrician responsibility. This demarcation point varies by utility but often occurs at the meter or service head.

Smart Grids

These modern grids use digital technology to monitor and control the distribution of electricity, improving efficiency and reliability. Smart grid systems represent the evolution of traditional power distribution into a more responsive, automated network.

Components of smart grid systems include:

- Advanced metering infrastructure (AMI) providing two-way communication
- Distribution automation equipment for remote switching and fault isolation
- Sensors throughout the system monitoring voltage, current, and power quality
- Data analytics systems identifying patterns, predicting failures, and optimizing performance

For electricians, smart grid technology introduces new considerations:

- Compatibility with demand response systems that may cycle loads
- Power quality monitoring equipment that records disturbances
- Integration with customer-owned generation (like solar) and energy storage
- More sophisticated protection devices with communication capabilities

As distribution systems evolve, electricians increasingly encounter smart devices that communicate with utility systems. Understanding how these components interact with traditional electrical systems will become increasingly important throughout your career.

The power distribution system directly impacts your work as an electrician—from service sizing and voltage considerations to troubleshooting outages and power quality issues. A solid grasp of these systems helps you communicate effectively with utility personnel and solve problems that cross the boundary between utility and customer systems.

4.4 Transformers, Generators, and Backup Power Systems

When a hospital loses utility power, lives hang in the balance. When a data center goes dark, millions in revenue can vanish in minutes. The systems that prevent these disasters—transformers, generators, and backup power equipment—represent some of the most critical components you'll encounter as an electrician.

These aren't just theoretical parts of the power system; they're the equipment you'll install, maintain, and troubleshoot throughout your career. From the pole-mounted transformer feeding a residential neighborhood to the massive generator system backing up a manufacturing plant, these systems form the backbone of reliable electrical service. Their proper installation and maintenance often mark the difference between minor inconvenience and catastrophic failure.

Transformers

Transformers are the workhorses of electrical distribution systems, appearing at multiple points between power generation and end use. They allow voltage to be stepped up for efficient transmission and stepped down for safe utilization—without these devices, our electrical grid simply couldn't function.

- **Purpose**: Transformers adjust voltage levels between the generation point and the end-user to minimize energy loss. They work on the principle of electromagnetic induction, transferring electrical energy between circuits through inductively coupled conductors.
- **Step-Up Transformers**: Increase voltage for efficient long-distance transmission. These massive transformers, found at generating stations, might take 24kV from a generator and step it up to 345kV or higher for transmission. Higher voltage means lower current for the same power, resulting in dramatically reduced line losses over distance.
- **Step-Down Transformers**: Decrease voltage to safe levels for use in homes and businesses. These transformers are found at substations and on utility poles, converting transmission or distribution voltages down to utilization voltages. The step-down process might occur in several stages—from transmission voltage to distribution voltage at substations, then further reduction at local distribution transformers.
- **Transformer Types**: Single-phase transformers for residential use and three-phase transformers for industrial applications. Single-phase transformers typically provide 120/240V service for homes, while commercial buildings often use three-phase transformers delivering 208Y/120V or 480Y/277V.

The NEC contains specific requirements for transformers in Article 450 that address installation clearances, overcurrent protection, grounding, and ventilation. When working with transformers, pay careful attention to the primary and secondary voltages, KVA ratings, impedance values, and connection types (delta or wye), as these factors determine how they're installed and protected.

Generators

Generators provide a lifeline when utility power fails or isn't available. Whether you're installing a small residential backup system or commissioning a massive generator for a hospital, understanding generator systems is essential for modern electrical work.

When a utility outage hits, generators create power through electromagnetic induction—the same principle used by power plants, just on a smaller scale. A prime mover (typically a diesel, natural gas, or gasoline engine) spins a rotor containing electromagnetic field coils inside a stator with conductive windings. This spinning magnetic field induces voltage in the stator windings, creating electrical power.

Types of Generators

- **Portable generators** (5-15kW) provide temporary power for construction sites, events, or emergency situations. As an electrician, you'll often need to set up temporary distribution systems from these units, ensure proper grounding according to NEC 250.34, and calculate maximum load capacity to prevent overloading.
- **Standby generators** (7-2000+ kW) provide automatic backup for homes and facilities. These permanently installed systems include automatic transfer switches, load control systems, monitoring equipment, and fuel systems sized for the expected outage duration..

4.5 Voltage Regulation and Load Calculations

Voltage regulation and load calculations are needed to ensure that electrical systems function within their safe operating limits. Proper voltage levels prevent damage to equipment, while load calculations ensure that circuits are neither overloaded nor undersized.

Voltage Regulation Definition

Voltage regulation measures how well a system maintains consistent voltage levels under varying load conditions. In an ideal system, voltage would remain perfectly stable regardless of load changes. In reality, voltage tends to drop as load increases and rise as load decreases.

The voltage regulation percentage is calculated as:

Regulation% = ((No-load Voltage—Full-load Voltage) ÷ Full-load Voltage) × 100%

Lower percentages indicate better regulation. Well-designed systems typically maintain regulation within 3-5%.

The Importance of Proper Voltage

Voltage outside acceptable ranges causes numerous problems:

- High voltage accelerates insulation breakdown and can destroy electronic components.
- Low voltage causes motors to draw excessive current, overheating and potentially failing.
- Voltage fluctuations can cause erratic operation of sensitive equipment like computers and medical devices.
- Inconsistent voltage leads to inconsistent performance of lighting and heating equipment.

ANSI C84.1 establishes standard voltage ranges:

- Service voltage: ±5% of nominal (114-126V for 120V systems)
- Utilization voltage: -13% to +6% of nominal (104-127V for 120V systems)

Methods of Voltage Regulation

Several methods help maintain proper voltage:

Automatic Voltage Regulators (AVRs) continuously monitor voltage and make adjustments to compensate for variations. These can be:

- Electronic regulators using solid-state components.
- Electromechanical regulators using variable transformers.
- Generator-mounted systems controlling field excitation.

Tap-Changing Transformers adjust the turns ratio to maintain secondary voltage as primary voltage varies. These come in two types:

- No-load tap changers (NLTC): Adjusted manually when the transformer is de-energized.
- Load tap changers (LTC): Can adjust automatically while carrying load.

Capacitor Banks help improve voltage regulation by reducing reactive power demand. These are commonly seen on distribution systems to support voltage during peak loads.

4.6 Load Calculations

Load calculations determine the amount of electrical demand on a circuit or system. These calculations are essential for properly sizing service equipment, feeders, and branch circuits. Accurate load calculations prevent both overloading (which creates safety hazards) and oversizing (which wastes money on unnecessary capacity).

Residential Load Calculations

Residential load calculations determine the minimum required service size for dwellings. These calculations follow NEC Article 220 Part III and involve several steps:

General Lighting Load

- Calculate the square footage of the dwelling
- Multiply by 3 VA per square foot (standard load)
- Include all general-use receptacles in this calculation

Small Appliance and Laundry Circuits

- Add 1,500 VA for each required small appliance circuit (minimum two)
- Add 1,500 VA for the laundry circuit

Fixed Appliances

- Add nameplate ratings for all fixed appliances (water heaters, ranges, ovens, dryers, etc.)
- Apply demand factors from NEC 220.53 for four or more appliances

HVAC Equipment

- Add 100% of the largest motor load
- Add 25% of each additional motor load

The total calculated load determines the minimum service size required. For a typical 3-bedroom home, this calculation often results in a 100A or 200A service requirement.

Commercial Load Calculations

Commercial load calculations are more complex due to the variety of equipment and usage patterns. These calculations follow NEC Article 220 Part IV and include:

Lighting Loads

- Calculate based on VA per square foot values from Table 220.12
- Apply demand factors from Table 220.42 for larger areas

Receptacle Loads

- General use: Calculate at 180VA per receptacle or 1VA per square foot
- Specific use: Calculate based on equipment requirements

Motor Loads

- Include 125% of the largest motor
- Add 100% of all other motors

HVAC Equipment

- Calculate based on actual equipment nameplate data

Commercial calculations often include additional considerations like power factor correction and harmonic mitigation for non-linear loads.

Factors Affecting Load Calculations

Several factors modify basic load calculations to reflect real-world usage patterns:

Demand Factor adjusts for the fact that not all connected loads operate simultaneously. For example, in a restaurant kitchen, not all cooking equipment runs at full capacity at the same time. The NEC provides specific demand tables for various load types.

- **Load Factor** represents the ratio of average load to peak load over time. Equipment with high load factors (like refrigeration) requires more robust electrical systems than equipment with low load factors (like seasonal HVAC).
- **Diversity Factor** accounts for differences in usage timing across multiple loads. This factor is particularly important in larger systems where statistical diversity can significantly reduce required capacity.
- **Future Expansion** considerations often add 25-30% additional capacity to commercial and industrial calculations to accommodate growth without requiring system upgrades.

Accurate load calculations ensure safe, code-compliant installations while preventing unnecessary costs from oversized equipment. As you prepare for your journeyman exam, focus on understanding the principles behind these calculations rather than just memorizing formulas.

4.6 Quiz-Testing Your Knowledge of Power Distribution and Sources

Now that you've learned the fundamentals of power generation, transformers, backup systems, and load calculations, it's time to test your understanding. This quiz will help you assess your grasp of the material and identify areas for further review.

Quiz Overview

- Multiple-choice, true/false, and calculation-based questions designed to test your knowledge of power distribution systems.
- Questions will cover the key concepts of generation, transformers, backup systems, voltage regulation, and load calculations.
- Detailed explanations for each answer will help reinforce learning.

Power Generation Questions

1. **Which power generation source typically provides the most consistent output regardless of weather conditions?**

    a) Solar power
    b) Wind power
    c) Nuclear power
    d) Hydroelectric power

2. **What is the primary advantage of renewable energy sources over conventional power generation?**

    a) They are always less expensive to install
    b) They have fewer environmental impacts
    c) They produce higher voltage output
    d) They require less maintenance

3. **True or False: Three-phase power is more efficient for motor operation than single-phase power.**

    a) True
    b) False

4. **Which of the following would NOT be classified as a renewable energy source?**
   a) Solar power
   b) Natural gas power plant
   c) Wind power
   d) Geothermal power

Transmission and Distribution Questions

5. **What is the primary reason for using extremely high voltages in transmission lines?**
   a) To increase the speed at which electricity travels
   b) To reduce line losses
   c) To reduce the cost of conductors
   d) To prevent birds from landing on the lines

6. **In a typical power distribution system, where are step-down transformers located?**
   a) At the power plant
   b) At substations
   c) Only inside buildings
   d) Only at generating stations

7. **Smart grid technology allows for:**
   a) Electricity to flow faster through the grid
   b) Two-way communication between utilities and consumers
   c) Elimination of all power outages
   d) Reduction in the cost of copper

8. **True or False: Underground distribution systems are less susceptible to weather damage but cost more to install and repair.**
   a) True
   b) False

Transformer Questions

9. **A transformer with 1,000 turns on the primary coil and 250 turns on the secondary coil has 480V applied to the primary. What is the secondary voltage?**

   a) 1,920V
   b) 120V
   c) 60V
   d) 240V

10. **Which transformer connection provides both phase-to-phase and phase-to-neutral voltages?**

   a) Delta
   b) Wye (Star)
   c) Corner-grounded delta
   d) Open delta

11. **True or False: The kVA rating of a transformer changes based on the connected load.**

   a) True
   b) False

12. **A 75 kVA transformer supplies a load with a 0.8 power factor. What is the maximum ampacity it can deliver at 208V, three-phase?**

   a) 208A
   b) 260A
   c) 360A
   d) 156A

Generator and Backup Power Questions

13. **According to NFPA 99, healthcare facilities' generators must be able to assume critical loads within:**

   a) 5 seconds
   b) 10 seconds
   c) 30 seconds
   d) 60 seconds

14. **True or False: When installing a backup generator, the neutral from the generator must always be bonded to ground regardless of transfer switch type.**

a) True
b) False

15. **Which of the following is NOT a typical component of a standby generator system?**

a) Automatic transfer switch
b) Battery starting system
c) Step-up transformer
d) Fuel storage system

16. **A 20kW standby generator is being installed for a residence. If the generator is rated at 83A at 240V, what size conductors (copper, THWN) are required between the generator and transfer switch?**

a) 6 AWG
b) 4 AWG
c) 3 AWG
d) 2 AWG

Voltage Regulation and Load Calculation Questions

17. **A transformer has a no-load voltage of 250V and a full-load voltage of 235V. What is its voltage regulation percentage?**

a) 6.38%
b) 15%
c) 6%
d) 10%

18. **True or False: Capacitor banks are often installed to improve voltage regulation by reducing reactive power demand.**

a) True
b) False

19. **When calculating the total load for a dwelling unit, small appliance circuits are added at:**

a) 1,000 VA each
b) 1,500 VA each
c) 2,000 VA each
d) 500 VA each

20. **A small office building has the following loads:

- Lighting: 12 kVA
- HVAC: 25 kVA
- Office equipment: 18 kVA
- Miscellaneous: 5 kVA

Using a demand factor of 0.8, what is the calculated demand load?**

a) 60 kVA
b) 48 kVA
c) 75 kVA
d) 45 kVA

21. **When calculating service size for motors, you must include what percentage of the largest motor?**

a) 100%
b) 125%
c) 150%
d) 80%

22. **A three-bedroom house has 2,000 square feet of living space. What is the minimum general lighting load per NEC?**

a) 2,000 VA
b) 3,000 VA
c) 6,000 VA
d) 4,000 VA

Answer Key and Explanations

1. **Answer: c) Nuclear power.** Nuclear power plants provide consistent baseload power regardless of weather conditions, unlike solar or wind which depend on environmental factors. Hydroelectric can be affected by drought conditions.

2. **Answer: b) They have fewer environmental impacts.** Renewable energy sources generally produce fewer greenhouse gases and pollution during operation compared to fossil fuel plants, though installation costs and maintenance requirements vary by technology.

3. **Answer: a) True.** Three-phase power provides more consistent torque for motors, is more efficient for power transmission, and allows for smaller conductors compared to single-phase power of the same voltage.

4. **Answer: b) Natural gas power plant.** Natural gas is a fossil fuel, not a renewable energy source. The other options (solar, wind, and geothermal) are all classified as renewable energy sources.

5. **Answer: b) To reduce line losses.** Higher voltages allow for lower current while transmitting the same amount of power. Since line losses are proportional to the square of current (I^2R), reducing current significantly reduces energy losses in transmission.

6. **Answer: b) At substations.** Substations contain transformers that step down transmission voltages to distribution levels. Additional step-down transformers may exist along the distribution system, but substations are the primary location.

7. **Answer: b) Two-way communication between utilities and consumers.** Smart grid technology enables utilities to receive information from consumers and not just send power, allowing for better load management, outage detection, and demand response programs.

8. **Answer: a) True.** Underground systems are protected from weather events like ice storms and high winds, but cost substantially more to install and repair due to excavation requirements.

9. **Answer: b) 120V.** Using the turns ratio formula: Vsecondary = Vprimary × (Nsecondary ÷ Nprimary), Vsecondary = 480V × (250 ÷ 1,000) = 480V × 0.25 = 120V

10. **Answer: b) Wye (Star).** Wye connections provide both phase-to-phase voltage (between any two phase conductors) and phase-to-neutral voltage (between any phase conductor and the neutral). Delta configurations only provide phase-to-phase voltage unless specially tapped.

11. **Answer: b) False.** A transformer's kVA rating is fixed by its design and cooling system. The connected load may be less than the transformer's rating, but the maximum capability doesn't change based on the load.

12. **Answer: b) 260A.** Using the formula: I = kVA ÷ (V × √3 × PF) for three-phase I = 75 kVA ÷ (0.208kV × 1.732 × 0.8) = 75 ÷ 0.288 = 260A

13. **Answer: b) 10 seconds** NFPA 99 requires emergency power to be available within 10 seconds for essential healthcare facility loads like life safety and critical care areas.

14. **Answer: b) False.** The neutral-ground bond should exist at only one point in the system. With a separately derived system, the bond is at the generator. With a non-separately derived system, the bond remains at the service equipment and must not be duplicated at the generator.

15. **Answer: c) Step-up transformer.** Standby generators typically produce power at utilization voltage (often 240V or 480V) and don't require step-up transformers. The other components (transfer switch, battery starter, and fuel storage) are standard parts of backup systems.

16. **Answer: b) 4 AWG**. Generator conductors must be sized at 115% of nameplate current per NEC 445.13. 83A × 1.15 = 95.45A According to Table 310.16, 4 AWG copper THWN has an ampacity of 95A at 75°C, making it the minimum size required.

17. **Answer: a) 6.38%.** Using the voltage regulation formula: Regulation% = ((No-load Voltage —Full-load Voltage) ÷ Full-load Voltage) × 100% Regulation% = ((250V —235V) ÷ 235V) × 100% = (15V ÷ 235V) × 100% = 6.38%

18. **Answer: a) True**. Capacitor banks provide reactive power compensation, which improves voltage regulation by reducing the reactive component of current in the system. This is especially beneficial for heavily inductive loads like motors.

19. **Answer: b) 1,500 VA each.** NEC 220.52(A) requires small appliance branch circuits to be calculated at 1,500 VA each, with a minimum of two such circuits required for the kitchen, pantry, breakfast room, and dining room areas.

20. **Answer: a) 60 kVA.** Total connected load = 12 + 25 + 18 + 5 = 60 kVA Applying the demand factor: 60 kVA × 0.8 = 48 kVA The answer is 48 kVA.

21. **Answer: b) 125%.** NEC 430.24 requires that the conductor that supplies a motor load must have an ampacity of not less than 125% of the motor's full-load current rating.

22. **Answer: c) 6,000 VA**. Per NEC 220.12, the general lighting load for dwelling units is calculated at 3 VA per square foot. 2,000 sq. ft. × 3 VA = 6,000 VA

Assessment

Count your correct answers:

- 18-22 correct: Excellent! You have a strong understanding of power distribution systems.
- 14-17 correct: Good job! Review the questions you missed to strengthen your knowledge.
- 10-13 correct: You're on the right track, but should review the chapter more thoroughly.
- Below 10: Take time to re-read the chapter and work through the examples again before moving forward.

5

Overcurrent Protection and Safety Devices

Electricity follows the path of least resistance—a principle that makes it both predictable and potentially dangerous. Like water finding its way through cracks in a dam, electrical current will exploit any weakness in a system, sometimes with spectacular and destructive results

In previous chapters, we explored the fundamental laws governing electrical flow and circuit design. Now, we turn our attention to one of the most critical aspects of electrical systems: protection. This isn't just about technical specifications—it's about preventing disasters, saving lives, and ensuring that the incredible power we've harnessed remains our servant rather than becoming our master.

Overcurrent protection devices are the unsung heroes of electrical systems. They stand guard 24/7, ready to react in milliseconds when things go wrong. From the humble fuse that sacrifices itself to save a circuit to sophisticated digital relays that can distinguish between normal surges and dangerous faults, these devices represent our best defense against electrical fires and equipment damage.

When Thomas Edison was developing his electrical systems, losses from overheating and fires were common and sometimes catastrophic. Today, we take for granted that plugging in a device won't burn down our homes—a confidence built entirely on the protective devices we'll explore in this chapter.

5.1 Circuit Breakers, Fuses, and Relays

Every burned conductor I've uncovered, every melted connection I've replaced, and every fire-damaged panel I've rewired tells the same story: electrical systems will eventually face abnormal conditions. It's not a matter of if, but when. Your choice of protection devices determines whether these events end with a simple reset or a catastrophic failure.

These protection devices—circuit breakers, fuses, and relays—form the critical backbone of electrical safety that professionals too often take for granted. I've watched apprentices dismiss them as mere "code requirements" until they witness firsthand the consequences of inadequate protection: equipment destroyed in seconds and, in worst cases, lives endangered.

Understanding the strengths, limitations, and appropriate applications of each protection type isn't just academic knowledge—it's practical wisdom that will serve you throughout your career. As we examine each option, I'll share not just the technical specifications, but the hard-earned lessons from decades of field experience that textbooks simply can't provide.

Circuit Breakers

Picture this: It's 3 AM, and suddenly your house goes dark. You stumble to the electrical panel, find the tripped breaker, and with a simple flip, power is restored. This everyday miracle is a circuit breaker in action—an ingenious device that has saved countless homes from electrical fires.

Purpose: Automatically interrupt the flow of electricity when a circuit becomes overloaded or experiences a short circuit.

How They Work: Inside that modest plastic housing lies a fascinating mechanism. When current exceeds safe levels, one of two things happens:

- In thermal breakers, a bimetallic strip heats up, bends, and triggers the switch
- In magnetic breakers, excessive current creates a magnetic field strong enough to pull a lever and break the circuit
- Combination breakers use both principles, providing fast response to short circuits and measured response to gradual overloads

Reset Functionality: Unlike fuses, circuit breakers can be reset after tripping, making them more convenient for long-term use. This reusability isn't just convenient—it saves money and reduces waste over time.

Applications: From the humble 15-amp breakers protecting your bedroom outlets to massive industrial units that can interrupt currents powerful enough to melt steel, circuit breakers scale to match nearly any electrical protection need. They're common in residential and commercial settings, protecting a variety of circuits.

Types of Circuit Breakers

Circuit breakers come in various designs, each optimized for specific applications and protection requirements:

Thermal Circuit Breakers

These workhorses of residential protection operate on a simple yet effective principle: as current flows through the breaker, it generates heat. When too much current flows, a bimetallic strip inside bends from the excess heat, triggering the mechanism that breaks the circuit. Think of them as the patient guardians of your home's electrical system—slow to anger but reliable when it matters.

Magnetic Circuit Breakers

When milliseconds matter, magnetic breakers shine. They use an electromagnetic coil that creates a magnetic field proportional to the current flowing through it. When current spikes dramatically—as in a short circuit—the magnetic field becomes strong enough to pull a mechanical latch, instantly breaking the circuit. These are the rapid responders of the circuit breaker family, ideal for protecting sensitive electronics and preventing arc flash hazards.

Thermal-Magnetic Circuit Breakers

These hybrid breakers use thermal protection for sustained overloads and magnetic tripping for sudden short circuits. This makes them versatile protectors suitable for a wide range of applications, from home circuit panels to industrial equipment. Their dual-protection approach is why they've become the standard choice in most modern electrical systems.

Electronic Circuit Breakers

The digital revolution has transformed circuit protection, too. Electronic breakers use microprocessors to monitor current with exceptional precision, allowing for customizable trip curves and remote monitoring capabilities. They can detect subtler problems that mechanical breakers might miss, making them ideal for critical infrastructure and sensitive equipment where failure isn't an option.

5.2 Fuses

Fuses are one of the oldest and most reliable forms of circuit protection, and date back to Thomas Edison's early electrical systems. Despite being gradually replaced by circuit breakers in many applications, fuses remain critical components in numerous electrical systems due to their unique advantages.

Purpose and Operation

Fuses protect circuits by melting a wire inside the fuse when the current exceeds a certain limit. This operation is based on a simple principle: controlled failure. The fuse element is designed to be the weakest link in the circuit, sacrificing itself to protect more valuable components downstream.

The fuse element heats up as current flows through it. Under normal conditions, this heat dissipates without damaging the element. When overcurrent occurs, the element heats faster than it can cool, causing it to melt and open the circuit. This characteristic makes fuses inherently reliable—they require no mechanical parts or external power to function.

Fuses respond to both short-duration high current faults (short circuits) and prolonged lower-level overcurrents that can cause overheating. Their time-current characteristics are precisely engineered for specific applications.

Types of Fuses

Cartridge fuses come in cylindrical shapes with contacts at each end. Common types include:

- Class H (formerly NEC standard).
- Class J (high-interrupting capacity, time-delay).
- Class K (replacement for Class H with higher interrupting ratings).
- Class L (high-current applications, 601-6000A).
- Class RK1 and RK5 (dual-element, time-delay).
- Class T (fast-acting, high-interrupting, compact size).
- Class CC (control circuits, high-interrupting).

Plug fuses screw into standard fuse sockets and are used in older residential installations. They include:

- Edison-base fuses (Type T).
- Type S fuses with adapters to prevent over-fusing.
- Time-delay versions for motor loads.

Semiconductor fuses are specially designed to protect electronic equipment and have extremely fast-acting characteristics measured in microseconds or milliseconds.

Resettable fuses (technically positive temperature coefficient thermistors) can return to service after cooling down. While convenient, they don't offer the precise protection of traditional fuses.

Advantages of Fuses

Fuses offer several benefits that make them preferable in certain applications:

- **Speed of operation**: Fuses can clear faults in as little as 1/4 cycle, faster than most circuit breakers.
- **No moving parts**: Higher reliability with less maintenance.
- **Higher interrupting ratings**: Many fuses can interrupt currents up to 300,000A.
- **Precise coordination**: Easier to achieve selective coordination in systems requiring multi-level protection.
- **Indication of failure**: Visual confirmation of operation (though not always visible).
- **No calibration needed**: Fuses don't drift out of calibration over time.
- **Lower arc-flash energy**: Faster clearing times reduce incident energy during faults.

Disadvantages of Fuses

Despite their advantages, fuses have limitations:

- **Single-use devices**: Once a fuse blows, it must be replaced, which is less convenient than resetting a circuit breaker.
- **Potential for improper replacement**: Nothing prevents installation of incorrect fuse types or ratings.
- **Single-phase operation**: In three-phase systems, a single fuse can open, creating dangerous single-phasing conditions for motors.
- **Stock requirements**: Facilities must maintain inventories of various fuse types and ratings.
- **Downtime during replacement**: Systems remain de-energized until fuses are replaced.

The NEC includes specific requirements for fuse installations, particularly regarding accessibility, grouping, and labeling. Proper fuse selection involves considering not just ampere rating but also voltage rating, interrupting capacity, time-current characteristics, and physical dimensions.

5.3 Relays

Relays are the decision-makers in electrical systems. They detect abnormal conditions and respond by opening or closing circuits. While fuses and circuit breakers provide basic overcurrent protection, relays offer sophisticated control based on multiple electrical parameters and conditions.

Types of Relays

Electromagnetic relays use a coil that gets magnetized when current flows through it. This magnetism attracts a movable armature, which opens or closes contacts. Common types include:

- Overcurrent relays that respond to excessive current.
- Undervoltage/overvoltage relays that detect abnormal voltage conditions.
- Differential relays that compare current entering and leaving a protected zone.
- Distance relays that respond to impedance changes indicating faults.
- Time-delay relays that introduce deliberate delays before operating.

Solid-state relays use semiconductor devices instead of mechanical components. These offer advantages including:

- No moving parts, resulting in longer operational life.
- Faster operation, often in milliseconds.
- No contact arcing or bounce.
- Ability to switch at the zero-crossing point, reducing electromagnetic interference.
- Compatibility with digital control systems.

Microprocessor-based relays represent the latest evolution in relay technology. These intelligent devices:

- Monitor multiple parameters simultaneously.
- Store fault data for later analysis.
- Communicate with supervisory control systems.
- Adapt their settings based on system conditions.
- Perform self-diagnostics to ensure reliable operation.

5.4 Short Circuit and Overload Protection

Overcurrent protection isn't just about preventing damage from high current—it's about distinguishing between short circuits and overloads, which require different responses. Understanding this distinction is essential for proper system design and could mean the difference between a minor inconvenience and a catastrophic failure.

Short Circuits

A short circuit occurs when a live wire comes into contact with a neutral or ground wire, causing a sudden surge in current.

Short circuits also represent one of the most dangerous electrical faults. When the normal circuit path is bypassed through accidental contact between conductors, resistance drops dramatically and current skyrockets—often to thousands of amps. This massive current spike occurs almost instantaneously and can generate tremendous heat and magnetic forces.

They come in several forms: phase-to-ground (current flows from a phase conductor to ground), phase-to-neutral (current flows from a phase conductor to the neutral), and phase-to-phase (current flows directly between phase conductors).

Protection Mechanisms: Circuit breakers and fuses are designed to detect and interrupt short circuits quickly to prevent fire or damage.

Effective short-circuit protection requires devices that can detect fault current rapidly (within milliseconds), interrupt the fault current safely without self-destruction, and withstand the tremendous mechanical forces during interruption.

Protection devices must have adequate interrupting capacity (AIC rating) to safely clear the maximum available fault current. Using a device with insufficient interrupting capacity can result in catastrophic failure during a fault.

The NEC requires equipment to be marked with its short-circuit current rating (SCCR) or AIR (Ampere Interrupting Rating) so electricians can verify it's appropriate for the installation location.

5.5 Overloads

Overloads occur when too many devices draw power from a circuit or when equipment malfunctions and draws more current than normal. While short circuits involve current levels 10-1000 times normal, overloads typically involve currents just 1.1-6 times the rated capacity. This difference in magnitude requires different protection responses.

Common causes of overloads include:

- Too many appliances operating simultaneously on a single circuit.
- Motor locked rotor conditions.
- Failed bearings causing mechanical binding.
- Excessive load on mechanical equipment.
- Voltage sags causing equipment to draw higher current.

Effect on the System

Overloads can cause overheating of wires and devices, which may lead to damage or fire if not addressed.

Unlike short circuits that cause immediate damage, overloads create problems over time through heat accumulation. The I^2R heating effect causes conductors and equipment to rise in temperature gradually. This heat breaks down insulation, increases resistance (creating more heat), and eventually can lead to insulation failure and fire.

Continuous overloads can:

- Reduce the life expectancy of motors and equipment.
- Damage insulation on conductors.
- Trip circuit breakers inconveniently.
- Create fire hazards in walls, ceilings, or equipment.
- Lead to efficiency losses from increased conductor resistance.

5.6 NEC Requirements for Overcurrent Protection

One of the most interesting shifts in the NEC over time has been the evolution of overcurrent protection requirements. What started as simple fusing rules has developed into comprehensive protection schemes that address everything from conductor protection to equipment safeguards. These requirements form the backbone of electrical safety in every installation you'll encounter.

The NEC addresses overcurrent protection primarily in Article 240, though requirements appear throughout the code. These provisions establish where protection is required, how it must be sized, and what types of protective devices are permitted for different applications.

5.7 NEC Compliance and Inspection

Successfully installing overcurrent protection isn't just about following code requirements—it's about demonstrating compliance to inspectors and ensuring long-term safety through proper maintenance and testing.

Ensuring Compliance

Installers must verify that overcurrent protection devices meet the specifications outlined in the NEC.

Documentation plays a crucial role in demonstrating compliance. For commercial and industrial installations, you'll often need to provide:

- Short-circuit current calculations showing adequate interrupting ratings.
- Coordination studies proving selective operation for critical systems.
- Equipment labeling showing available fault current at service equipment per NEC 110.24.
- Load calculations justifying the sizing of protective devices.

Proper installation is equally important. This includes:

- Maintaining required working clearances around equipment (NEC 110.26)
- Properly torquing connections to manufacturer specifications (NEC 110.14(D))
- Installing the correct breaker types for the panel being used
- Following tap rules precisely when conductors aren't protected at their supply point
- Using properly rated equipment for the environment (wet locations, corrosive atmospheres, etc.)

Inspection Procedures

The code outlines procedures for testing and inspecting overcurrent devices to ensure they are functioning correctly.

Initial inspection by the Authority Having Jurisdiction (AHJ) focuses on verifying:

- Proper sizing of overcurrent devices relative to conductor capacity
- Correct installation according to manufacturer instructions
- Appropriate device ratings for the installation environment
- Adequate interrupting capacity for available fault current
- Proper labeling of circuits and equipment

Beyond initial inspection, NFPA 70B (Recommended Practice for Electrical Equipment Maintenance) suggests regular maintenance including:

- Thermal scanning to detect hot spots before failure.
- Operating circuit breakers periodically to ensure mechanism functionality.
- Testing trip units in larger breakers to verify proper operation.
- Checking connections for proper torque.
- Visually inspecting for signs of damage or deterioration.

A growing number of jurisdictions are requiring documented maintenance of overcurrent protective devices in commercial and industrial facilities. This shift recognizes that even properly installed protection can fail if not maintained.

Compliance with overcurrent protection requirements is perhaps the most fundamental aspect of electrical safety. While other code violations might reduce efficiency or convenience, inadequate overcurrent protection directly threatens lives and property.

5.8 Quiz: Testing Your Knowledge of Overcurrent Protection and Safety Devices

Now that you've learned about circuit breakers, fuses, relays, short circuits, overloads, and the NEC requirements for overcurrent protection, it's time to test your understanding. This quiz will help reinforce the concepts you've learned and identify areas where you may need further review.

Practice Questions

1. **Which of the following is a primary difference between short circuits and overloads?**

   a) Short circuits involve gradual current increases, while overloads are sudden
   b) Short circuits involve sudden, extreme current increases, while overloads are gradual
   c) Short circuits affect only three-phase systems
   d) Overloads only occur in residential applications

2. **According to NEC 240.4(D), what is the maximum overcurrent protection permitted for a 12 AWG copper conductor with 90°C insulation?**

   a) 15 amperes
   b) 20 amperes
   c) 25 amperes
   d) 30 amperes

3. **What is the primary advantage of a Class RK1 fuse over a standard Class H fuse?**

   a) Lower cost
   b) Higher interrupting rating
   c) Ability to be reset after operation
   d) Lower voltage drop

4. **True or False: The interrupting rating of an overcurrent device must be at least equal to the maximum available fault current at its installation point.**

   a) True
   b) False

5. **Which type of circuit breaker uses both thermal and magnetic trip mechanisms?**

   a) GFCI breaker
   b) Thermal-magnetic breaker
   c) Solid-state trip breaker
   d) AFCI breaker

6. **In a commercial building, a 480V circuit feeds a motor load. Which of the following would be correctly sized for short-circuit and ground-fault protection of a motor with a full-load current rating of 28A?**

   a) 28A circuit breaker
   b) 30A circuit breaker
   c) 40A circuit breaker
   d) 70A circuit breaker

7. **What is the primary purpose of selective coordination in an overcurrent protection system?**

   a) To ensure all devices trip simultaneously for maximum protection
   b) To ensure only the device closest to the fault operates
   c) To reduce voltage drop in the system
   d) To allow for future expansion of the system

8. **Which of the following is NOT a type of relay used for overcurrent protection?**

   a) Differential relay
   b) Deceleration relay
   c) Instantaneous overcurrent relay
   d) Time-delay overcurrent relay

9. **True or False: Fuses operate more quickly than circuit breakers for the same level of overcurrent.**

   a) True
   b) False

10. **According to the NEC, where should overcurrent protection devices typically be located?**

    a) At the load end of each circuit
    b) At the point where conductors receive their supply
    c) Only at the service entrance
    d) Every 100 feet along the circuit length

11. **Which of the following is a proper application of NEC tap rules?**

    a) A 50-foot tap conductor is protected only at its supply end
    b) A 25-foot tap conductor sized at the same ampacity as the feeder is protected only at its load end
    c) A 10-foot tap conductor is protected at its load end but not at its supply end
    d) A 15-foot tap conductor sized at least 1/3 the ampacity of the feeder is protected at its load end

12. **What type of overcurrent protection is specifically designed to protect against arcing faults that standard breakers might not detect?**

    a) GFCI
    b) AFCI
    c) Differential relay
    d) Thermal-magnetic breaker

13. **What is the maximum setting allowed for motor overload protection according to NEC 430.32 for a motor with a service factor of 1.15 or more?**

    a) 100% of full-load current
    b) 115% of full-load current
    c) 125% of full-load current
    d) 140% of full-load current

14. **Which of the following would NOT be considered an overload condition?**

    a) Too many appliances operating on a single circuit
    b) A motor with a locked rotor
    c) A direct short between phase conductors
    d) Reduced ventilation causing motor overheating

15. **What does the "time-current characteristic curve" of an overcurrent device show?**

    a) How much time must elapse before a device can be reset
    b) How long the device will last before needing replacement
    c) How quickly the device will operate at different levels of overcurrent
    d) How the device's temperature changes over time

Answer Key and Explanations

1. **Answer: b) Short circuits involve sudden, extreme current increases, while overloads are gradual**

   Short circuits cause an immediate spike in current, often 10-1000 times normal levels, due to the drastically reduced resistance in the circuit path. Overloads typically involve currents just 1.1-6 times normal and build up gradually, causing heating over time.

2. **Answer: b) 20 amperes**

   NEC 240.4(D) specifically limits 12 AWG copper conductors to 20A protection, regardless of the temperature rating of the insulation. This is one of the "small conductor rules" that provides added safety for common wire sizes.

3. **Answer: b) Higher interrupting rating**

   Class RK1 fuses have interrupting ratings up to 200,000A, while standard Class H fuses are limited to 10,000A. RK1 fuses also offer time-delay characteristics and better current limitation, but their higher interrupting capacity is the primary advantage.

4. **Answer: a) True**

   NEC 110.9 requires that equipment intended to interrupt current at fault levels have an interrupting rating sufficient for the current that must be interrupted. Using a device with inadequate interrupting capacity creates a significant safety hazard.

5. **Answer: b) Thermal-magnetic breaker**

   Thermal-magnetic breakers use a bimetallic element for overload protection (thermal) and an electromagnetic mechanism for short-circuit protection (magnetic). This combination provides both time-delayed and instantaneous protection.

6. **Answer: d) 70A circuit breaker**

   NEC 430.52 allows up to 250% of full-load current for inverse time breakers protecting motor circuits. $28A \times 2.5 = 70A$. This higher setting prevents nuisance tripping during motor starting while still providing short-circuit protection.

7. **Answer: b) To ensure only the device closest to the fault operates**

   Selective coordination ensures that only the protective device immediately upstream from a fault opens, isolating only the affected circuit while maintaining power to unaffected portions of the system. This is particularly important in critical facilities like hospitals.

8. **Answer: b) Deceleration relay**

   This is a fictional type of relay. The other options are legitimate protective relay types used in power systems. Differential relays compare current entering and leaving a zone, while instantaneous and time-delay overcurrent relays respond to current exceeding set thresholds.

9. **Answer: a) True**

   Fuses generally operate more quickly than circuit breakers for the same level of overcurrent. Current-limiting fuses can clear faults in as little as 1/4 cycle (about 4 milliseconds), while circuit breakers typically require a minimum of 1 cycle (about 16.7 milliseconds).

10. **Answer: b) At the point where conductors receive their supply**

    NEC 240.21 requires that overcurrent protection be provided at the point where conductors receive their supply. This ensures that the entire length of the conductor is protected. The code does allow certain exceptions under specific tap rules.

11. **Answer: d) A 15-foot tap conductor sized at least 1/3 the ampacity of the feeder is protected at its load end**

    NEC 240.21(B)(2) allows 10-foot tap conductors to be protected only at their load end if they meet specific requirements, including being at least 1/3 the ampacity of the supply conductors. A 15-foot tap would comply with this rule.

12. **Answer: b) AFCI**

    Arc-Fault Circuit Interrupters (AFCI) are specifically designed to detect and interrupt arcing faults that may not draw enough current to trip standard overcurrent protection but still present a fire hazard.

13. **Answer: c) 125% of full-load current**

    NEC 430.32(A)(1) permits motors with a service factor of 1.15 or greater to have overload protection set at no more than 125% of the motor's full-load current rating.

14. **Answer: c) A direct short between phase conductors**

    A direct short between phase conductors is a short circuit, not an overload. The other options describe situations where current exceeds normal operating levels but remains within the range typically associated with overload conditions.

15. **Answer: c) How quickly the device will operate at different levels of overcurrent**

Time-current characteristic curves show the relationship between the amount of overcurrent and the time it takes for the device to operate. These curves are essential for coordinating protective devices and ensuring proper protection.

Assessment

Count your correct answers:

- 13-15 correct: Excellent! You have a solid understanding of overcurrent protection.
- 10-12 correct: Good job! Review the questions you missed.
- 7-9 correct: You're on the right track, but you should review the chapter more thoroughly.
- Below 7: Take time to re-read the chapter and work through the examples again.

Understanding overcurrent protection is fundamental to electrical safety. These concepts will appear throughout the exam and form the basis for many decisions you'll make throughout your career as a journeyman electrician.

6

Electrical Components and Devices

The manufacturing plant had been down for three hours when I arrived. A critical control panel had failed, halting production and costing thousands in lost revenue every minute. The maintenance team had already replaced a circuit breaker, two relays, and even swapped in a spare PLC—all to no avail.

"We've changed everything we can think of," the plant manager said, frustration evident in his voice.

I opened the panel and began methodically checking each component, starting with the power supply. The input voltage measured correctly, but the output showed nothing. Tracing the circuit further, I found a small terminal block with a loose connection. This seemingly insignificant component—just a piece of plastic with some metal screws—had shut down an entire factory.

Twenty minutes later, production resumed. The manager was amazed that something so simple could cause such a massive problem.

"It's always the components nobody thinks about," I told him. "The fancy electronics get all the attention, but it's often the basic parts that make or break a system."

This truth has proven itself countless times throughout my career. Understanding electrical components—from the simplest terminal block to the most sophisticated programmable controller—is what separates electricians who can truly troubleshoot from those who just swap parts, hoping for the best.

As electricians, we work with a vast array of devices and components daily. Each has specific functions, installation requirements, and failure modes. Recognizing how these pieces work together in a system allows us to design better installations, troubleshoot more effectively, and ensure safety and reliability.

In this chapter, we'll examine the most common electrical components and devices you'll encounter as a journeyman electrician. We'll cover their functions, applications, and the code requirements that govern their installation. Whether you're wiring a new building or troubleshooting an existing system, this knowledge will help you work more effectively and pass your licensing exam with confidence.

6.1 Conductors

Conductors are materials that allow electricity to flow through them easily.

Every circuit you'll ever install, troubleshoot, or maintain relies on conductors to carry current where it's needed. These materials contain loosely bound electrons in their atomic structure, allowing electrons to move freely when voltage is applied. This movement of electrons is what we know as electrical current.

Common Conductors

Copper and aluminum are the most commonly used conductors in electrical systems due to their excellent conductivity.

Copper has been the standard in electrical work for generations, offering excellent conductivity, good mechanical strength, and corrosion resistance. With a conductivity rating of 100% on the International Annealed Copper Standard (IACS), copper serves as the benchmark against which other conductors are measured. It's commonly used in:

- branch circuit wiring
- feeders and service conductors
- motor windings
- grounding systems
- busbars in distribution equipment

Aluminum offers about 61% of copper's conductivity but weighs only about 30% as much. This makes it particularly valuable for overhead transmission lines where weight matters. Aluminum is also significantly less expensive than copper, making it economical for large-sized conductors.

Modern aluminum alloys have overcome many of the problems that gave aluminum a bad reputation in the 1960s and 70s. Aluminum is typically used in:

- Utility transmission and distribution lines.
- Large service entrance conductors (usually 4/0 and larger).
- Some feeder conductors.
- Transformer windings.
- Utility-grade busbars.

Other conductive materials you might encounter include:

- Silver (106% IACS, used in specialized electronics and high-end connections).
- Gold (70% IACS, used in corrosion-resistant connections).
- Brass (28% IACS, used in some terminals and connectors).
- Steel (used where mechanical strength is paramount, often as support in ACSR conductors).

Applications

Wires, cables, and electrical connectors rely on conductors to transmit electrical current efficiently.

The specific conductor you choose depends on the application's requirements:

- For residential branch circuits, copper is standard due to its reliability and ease of installation.
- For large feeders and services, aluminum may be more economical.
- For grounding, copper is preferred due to its corrosion resistance.
- For specialized applications like motor windings, high-purity copper may be specified.

The size of the conductor must match the current-carrying requirements of the circuit. The NEC provides ampacity tables (primarily in Article 310) that specify the allowable current for different conductor sizes based on installation method, insulation type, and ambient conditions.

Resistance

The lower the resistance of a conductor, the more efficiently it carries current.

All conductors have some resistance to current flow, which causes voltage drop and heat generation. This resistance depends on several factors:

- **Conductor material**: Different materials have different inherent resistivity. Copper has lower resistivity than aluminum, which is why a copper conductor can carry more current than an aluminum conductor of the same size.
- **Conductor size**: Larger conductors have lower resistance. Each increase in AWG size

reduces resistance by approximately 20%.

- **Temperature**: As conductors heat up, their resistance increases, which can lead to a spiral effect of more heating and even higher resistance. This is why ampacity tables include different columns for different temperature ratings.
- **Length**: Resistance increases linearly with length. A 200-foot run has twice the resistance of a 100-foot run of the same conductor.

6.2 Semiconductors

Semiconductors have properties between conductors and insulators and are essential for controlling electrical current in devices such as transistors.

While traditional electrical work focuses primarily on conductors and insulators, the modern electrician increasingly encounters semiconductor-based devices. These remarkable materials bridge the gap between conductors and insulators, with electrical properties that can be precisely controlled. Their unique ability to sometimes conduct and sometimes insulate has revolutionized our electrical world.

Common Semiconductors

Silicon and germanium are the most widely used materials for semiconductors.

Silicon dominates the semiconductor industry, accounting for over 95% of all semiconductor devices. Its abundance in the earth's crust (it's the second most common element after oxygen) makes it relatively inexpensive, while its stable properties make it ideal for most applications. Silicon is used in:

- Computer processors and memory
- Power control devices like SCRs and triacs
- Photovoltaic cells
- Rectifiers and diodes
- Voltage regulators

Germanium was used in early semiconductor devices and still finds applications in specialized areas due to its superior performance at high frequencies and low forward voltage drop. You might encounter germanium in:

- High-speed communication circuits
- Some specialized diodes
- Infrared optics
- Certain photovoltaic applications

Gallium arsenide and other compound semiconductors offer performance advantages for specific applications:

- High-efficiency solar cells
- High-brightness LEDs
- Radio frequency devices
- Optoelectronic devices

Applications

Semiconductors are the backbone of modern electronics, including computers, solar cells, and LED lights.

As an electrician, you'll regularly work with semiconductor devices even if you don't directly handle the raw materials. These include:

Diodes allow current to flow in only one direction. They're used in:

- Power supplies for rectification (converting AC to DC)
- Surge protection devices
- LED lighting
- Solar panel blocking diodes

Transistors can amplify signals or act as switches. In electrical systems, you'll find them in:

- Variable frequency drives for motors
- Dimming controls
- Soft-start motor circuits
- Electronic ballasts for lighting

Thyristors (SCRs, triacs, and DIACs) control when and how much current flows. These devices appear in:

- Motor speed controls
- Light dimmers
- Heating element controls
- Phase control circuits

Integrated circuits combine multiple semiconductor components into a single package, performing complex functions. These appear in:

- Programmable timers
- Protection relays
- Smart controls for HVAC and lighting
- Energy monitoring systems

6.3 Switches, Outlets, and Lighting Controls

The humble light switch and receptacle outlet might seem like the simplest components we work with, but don't be fooled by their apparent simplicity. These devices form the primary interface between electrical systems and the people who use them. While a homeowner might view them as mere convenience points, experienced electricians recognize them as critical components that require proper selection, installation, and maintenance.

From basic toggle switches to sophisticated smart home controllers, from standard duplex receptacles to specialized hospital-grade outlets, these devices reflect the evolution of our electrical world. Their proper selection and installation require knowledge of load requirements, environmental conditions, code specifications, and user needs. As we'll see, there's much more to these everyday components than meets the eye.

6.4 Switches

A switch is a device used to open or close an electrical circuit. The basic concept seems simple: establish or interrupt the flow of current with the flip of a lever or push of a button. However, switches come in numerous configurations designed for specific applications. Selecting the right switch for each situation prevents operating problems, ensures user satisfaction, and maintains safety compliance.

Types of Switches

Single-pole, double-pole, three-way, four-way, and dimmer switches.

- **Single-pole switches** control a circuit from one location and have ON/OFF markings on the toggle. They have two brass terminals for the hot (ungrounded) conductor, which the switch connects or disconnects. These are the most common residential switches, used for controlling lights or receptacles from a single location.
- **Double-pole switches** control 240-volt equipment or simultaneously control multiple 120-volt circuits. They feature four brass terminals to switch both hot conductors in a 240-volt circuit. These switches are larger than single-pole types and often have a more substantial toggle or handle to accommodate the higher voltage and current.
- **Three-way switches** allow control of a circuit from two different locations. Recognizable by their lack of ON/OFF markings, they have three terminals: one common and two travelers. Three-way switches are commonly used at the top and bottom of staircases or opposite ends of hallways.
- **Four-way switches** work in conjunction with three-way switches to control a circuit from three or more locations. They have four terminals for the traveler wires and are installed between two three-way switches. For control from more than three locations, multiple four-way switches can be installed between the three-way switches.
- **Dimmer switches** adjust the brightness of connected lighting by varying the voltage supplied to the fixture. Modern dimmers use solid-state electronics rather than variable resistors, making them more efficient. When installing dimmers, you must ensure they're compatible with the type of lighting being controlled—particularly important with LED lighting.

Installation

Proper installation is critical for ensuring safe and reliable operation.

When installing switches:

- Connect the hot (ungrounded) conductor to the switch, never the neutral.
- Ensure proper grounding of metal switch boxes and switch yokes.
- Maintain proper wire connections —either wrapped clockwise around terminal screws or inserted into push-in terminals according to manufacturer instructions.
- Observe proper depth in the box to prevent pinched wires.
- Ensure adequate wire length (minimum 6 inches from box entry) per NEC 300.14.
- Follow manufacturer torque specifications when tightening terminal screws.

Common installation errors include:

- Improper identification of travelers in three-way switch installations.
- Overloading dimmers by connecting too many fixtures.
- Installing standard switches in locations requiring weather-resistant types.
- Failing to properly ground metal boxes or switch yokes.
- Using switches without the appropriate ampere or horsepower rating.

The NEC addresses switch installation in Article 404, and covers mounting, grounding, enclosures, and accessibility requirements. For the journeyman exam, pay particular attention to the provisions for grounding and the specific requirements for different switch types.

6.5 Outlets

An outlet, or receptacle, is a point in an electrical system where electrical devices are plugged in. Receptacles represent the end point of most branch circuits and the interface through which users access electrical power. Unlike hardwired connections, receptacles allow flexible use of electricity through plug-connected equipment. While they might seem like simple devices, receptacles involve critical safety features and must be selected and installed with care.

Types of Outlets

Standard outlets, GFCI (Ground Fault Circuit Interrupter) outlets, AFCI (Arc Fault Circuit Interrupter) outlets, and USB outlets.

Standard receptacles are rated for specific voltages and ampacities:

- 125V, 15A or 20A receptacles for standard household circuits (NEMA 5-15R and 5-20R).
- 250V receptacles for larger appliances like dryers and ranges (NEMA 6-20R, 14-30R, 14-50R).
- 125/250V receptacles for specialized equipment (NEMA 10-30R, 10-50R, 14-30R).

GFCI receptacles detect ground faults (current leakage to ground) and quickly disconnect power to prevent shock hazards. The NEC requires GFCI protection in locations with increased shock risk, including:

- bathrooms
- kitchens (countertop receptacles)
- outdoor locations
- garages
- crawl spaces
- Unfinished basements
- areas near sinks, tubs, or pools

AFCI receptacles detect dangerous arc faults that could lead to fires. These devices monitor the current waveform for patterns indicating arc faults and disconnect power when hazardous arcs are detected. The NEC requires AFCI protection in most habitable rooms of dwellings.

USB receptacles combine standard electrical outlets with built-in USB charging ports, eliminating the need for adapter plugs. These come in various configurations, including:

- Dual USB-A ports
- Combined USB-A and USB-C ports
- USB-C power delivery ports for faster charging

Specialty receptacles include:

- Hospital-grade receptacles (identified by a green dot) built to higher durability standards.
- Weather-resistant receptacles (marked “WR”) for outdoor use.
- Tamper-resistant receptacles (marked “TR”) with internal shutters to prevent objects from being inserted.
- Isolated ground receptacles (orange, with a triangle symbol) that separate the grounding path from the box and conduit system.

6.6 Lighting Controls

Lighting controls manage the power supplied to lighting fixtures. The world of lighting controls has evolved dramatically from simple toggle switches to sophisticated systems that respond to occupancy, daylight levels, time schedules, and even smartphone commands. These advanced control systems save energy, enhance comfort, and provide unprecedented flexibility in how spaces are illuminated.

Types of Lighting Controls

Manual switches, motion sensors, timers, and smart lighting systems.

Manual Switches are still the most basic form of lighting control, allowing users to turn lights on and off directly. Standard toggle and rocker switches handle basic on/off functionality, while dimmers add the ability to adjust light intensity. For control from multiple locations, three-way and four-way switches enable operation from different entrances to a space. More sophisticated options include scene controllers that activate preset lighting configurations with a single touch, and key switches that restrict access in security-sensitive areas.

Occupancy and Vacancy Sensors automatically control lighting based on detecting presence in a space. Passive infrared (PIR) sensors detect body heat movement, while ultrasonic sensors use sound wave reflections to identify motion. For critical applications, dual technology sensors combine multiple detection methods to prevent false triggers or missed occupancy. The distinction between occupancy and vacancy sensors is important: occupancy sensors both turn lights on when someone enters and off when they leave, while vacancy sensors require manual activation but turn off automatically when the space becomes vacant.

Daylight Controls adjust artificial lighting based on available natural light, maximizing energy savings while maintaining appropriate illumination levels. Simple photocells detect ambient light levels and switch lighting accordingly, while more sophisticated daylight harvesting systems gradually dim electric lights as natural light increases. These systems work particularly well in spaces with large windows, skylights, or other sources of natural illumination.

Time-Based Controls operate lighting on predetermined schedules, ideal for spaces with predictable usage patterns. Simple timers turn lights on and off at set times, while astronomical time clocks adjust for seasonal daylight changes automatically. More advanced systems offer programmable schedules for different days of the week and holiday scheduling capabilities to accommodate irregular operations.

Smart Lighting Systems integrate with building or home automation systems to provide enhanced functionality and remote control capabilities. These systems typically feature wireless controls accessible via smartphones and tablets, voice-activated operation through digital assistants, and integration with other building systems like HVAC and security. Many also collect usage data for energy analysis and optimization.

6.7 Motors, Relays, and Control Circuits

Motors, relays, and control circuits are the workhorses of industrial electricity. While lighting and receptacles might be more visible, these motor control components power the machinery that drives our modern world. From simple conveyor belts to sophisticated manufacturing equipment, motor systems convert electrical energy into the mechanical motion that makes industry possible.

These systems combine power circuits that deliver energy to motors with control circuits that direct when and how they operate. The seemingly simple act of starting a motor often involves multiple components working in precise sequence. When these systems falter, production stops—making motor control knowledge one of the most valuable skills in an electrician's toolkit.

6.8 Motors

A motor is a device that converts electrical energy into mechanical energy. The principle behind all electric motors is simple: when current flows through a conductor within a magnetic field, a force is produced. This electromagnetic interaction creates rotational movement that can drive machinery of all types. While the fundamental principle remains the same, motors come in countless variations designed for specific applications and operating conditions.

Types of Motors

AC (Alternating Current) motors and DC (Direct Current) motors.

AC Motors: operate directly from alternating current power sources and are the most common type in industrial and commercial applications.

Induction Motors: (asynchronous) feature a stator with windings that create a rotating magnetic field and a rotor that turns slightly slower than the field rotation (thus "asynchronous"). They are valued for their simplicity, durability, and relatively low cost. Common types include:

- Three-phase squirrel cage motors for industrial applications.
- Single-phase split-phase motors for smaller equipment
- Capacitor-start motors for applications requiring higher starting torque
- Shaded-pole motors for small fans and simple applications

Synchronous Motor*s:* maintain exact speed relative to line frequency, making them ideal for applications requiring precise timing. These motors either use permanent magnets or DC-excited rotors to lock into step with the rotating magnetic field.

DC Motors: offer excellent speed control and high starting torque, though they typically require additional components to operate from standard AC power. Common types include:

- Shunt-wound motors that maintain relatively constant speed under varying loads.
- Series-wound motors that deliver high starting torque.
- Compound motors that combine characteristics of both series and shunt types.
- Permanent magnet motors that eliminate field windings.
- Brushless DC motors that use electronic commutation instead of brushes.

Special Motor Types: serve specific applications:

- Servo motors for precise positioning and control.
- Stepper motors that rotate in discrete steps for accurate positioning.
- Variable frequency drive (VFD) systems provide adjustable speed control.
- Explosion-proof motors for hazardous environments.
- Submersible motors for underwater applications.

Motor Control

Motor control circuits regulate the speed, direction, and power of motors using contactors, relays, and other devices.

Basic motor control involves:

- Starting and stopping functions (manual or automatic)
- Speed regulation (fixed or variable)
- Direction control (forward/reverse operation)
- Protection against overloads and fault conditions

Control devices include:

- Magnetic contactors that handle the motor's main power
- Overload relays that protect against excessive current
- Control relays that sequence operations
- Timers that manage delay functions
- Limit switches that detect position
- Photoeyes and proximity sensors that detect presence

Modern motor control increasingly incorporates electronic devices:

- Variable frequency drives (VFDs) that provide precise speed control
- Soft starters that reduce inrush current during starting
- Programmable logic controllers (PLCs) that manage complex control sequences
- Motor protection relays that monitor multiple parameters
- Network interfaces that enable remote monitoring and control

The NEC addresses motor installation and protection primarily in Article 430, covering conductor sizing, overcurrent protection, disconnecting means, and controller requirements. For the journeyman exam, focus on motor circuit calculations, overload protection sizing, and the distinctions between branch circuit protection and running overload protection.

6.9 Relays

A relay is an electrically operated switch used to control circuits, particularly in automation and control systems.

At its core, a relay allows a small control signal to manage a much larger load circuit. This capability makes relays the bridge between control logic and power delivery in countless electrical systems. The basic electromagnetic relay has remained largely unchanged for decades, while newer solid-state versions offer enhanced features for modern applications.

6.10 Types of Relays

Electromagnetic Relays

Electromagnetic relays use an electromagnet to physically move contacts. When the control coil is energized, it creates a magnetic field that pulls the armature, changing the state of the contacts. These relays are valued for their simplicity, reliability, and complete isolation between control and load circuits.

Common types include:

- **Control relays** feature multiple contact arrangements for general control applications. They typically have several sets of contacts that can be normally open (NO), normally closed (NC), or combinations of both. Contact ratings vary from light-duty signal switching to substantial power loads.
- **Power relays** are designed to handle larger currents. These include contactors for motor

starting and lighting loads, which essentially function as heavy-duty relays designed for frequent operation under load.
- **Timing relays** incorporate delay functions. They can provide on-delay (delay after coil energization), off-delay (delay after coil de-energization), or more complex timing sequences for process control.

Solid-State Relays (SSRs)

Solid-State Relays use semiconductor components instead of mechanical contacts to switch circuits. These relays offer advantages including no moving parts, silent operation, no contact bounce, and potentially longer operational life.

SSRs typically employ:

- Triacs or SCRs for AC load switching, enabling features like zero-crossing switching that reduce electrical noise
- MOSFETs or IGBTs for DC load switching, providing low on-state resistance and efficient operation
- Optical isolation between input and output for safety and noise immunity

Thermal Relays

Thermal relays operate based on heat generated by current flow. The most common application is overload protection for motors, where the relay contains a heater element and bimetallic strips that respond to excessive current by opening contacts.

Specialty Relays

Several types of relays serve specific purposes:

- **Protective relays** detect abnormal conditions in power systems and initiate corrective action.
- **Reed relays** use magnetically activated reed switches for high-speed, low-power applications.
- **Latching relays** maintain their contact position without continuous coil power.
- **Safety relays** provide redundant, monitored switching for critical safety functions.

Applications

Relays are often used in motor control circuits, as well as in systems that require remote or automatic control of electrical devices.

In motor control systems, relays serve multiple functions:

- Power switching via contactors to start and stop motors
- Overload protection through thermal overload relays
- Sequencing operations through control relays
- Reversing motor direction via forward/reverse contactors
- Implementing safety interlocks to prevent dangerous operations

Building automation relies heavily on relays for controlling:

- Lighting systems, particularly in commercial buildings
- HVAC equipment, including fans, pumps, and compressors
- Access control and security systems
- Fire alarm response functions

Industrial automation uses relays for:

- Implementing logic functions in older or simpler control systems
- Interfacing between electronic controls and power circuits
- Emergency stop functions and safety systems
- Isolating different voltage systems from each other

6.11 Control Circuits

Control circuits are used to manage and regulate the operation of electrical systems and devices.

While power circuits deliver energy to loads like motors and heaters, control circuits determine when and how these loads operate. Control circuits typically operate at lower voltages and currents than the equipment they control, providing safer interaction for operators while enabling complex automated functions.

Components of Control Circuits

Control circuits may include switches, sensors, relays, contactors, and programmable logic controllers (PLCs).

- **Input Devices** gather information about conditions and operator intentions. Pushbuttons provide the most basic human interface, allowing operators to initiate actions like starting or stopping equipment. Selector switches enable operators to choose between different operating modes, while limit switches detect when moving parts reach specific positions. Various sensors monitor conditions like temperature, pressure, flow, and proximity without requiring physical contact.

- **Processing Devices** evaluate inputs and determine appropriate actions. Simple control systems might use relay logic, where the arrangement of contacts creates basic decision-making capability. Timers add time-based functions, allowing equipment to operate for preset durations or delay operations after specific events. More complex systems employ Programmable Logic Controllers (PLCs), which use software to implement sophisticated control strategies that would require extensive hardware in traditional relay systems.
- **Output Devices** execute the decisions to control equipment. Contactors serve as the workhorses of industrial control, switching power to motors and other high-current loads. Solenoids convert electrical signals to linear mechanical motion for valves and actuators. Indicator lights, displays, and audible alarms provide feedback to operators about system status and conditions requiring attention.

6.12 Quiz: Testing Your Knowledge of Electrical Components and Devices

They say the devil is in the details, and nowhere is this more true than in electrical work. A misidentified component, an incorrectly sized conductor, or a poorly selected control device can mean the difference between a system that works flawlessly for decades and one that fails catastrophically on day one.

Let's put your component knowledge to the test with questions that mirror real-world situations you'll face in the field. These aren't just academic exercises—they represent the kinds of decisions you'll make daily as a journeyman electrician. Each correct answer builds confidence that you can select, install, and troubleshoot the crucial components that make electrical systems work.

Ready to test your knowledge? Let's see how well you understand the building blocks of electrical systems.

Practice Questions

1. **Which of the following materials is the best electrical conductor?**

   a) Aluminum
   b) Copper
   c) Silver
   d) Gold

2. **What is the primary function of an insulator in an electrical system?**

   a) To allow free movement of electrons
   b) To prevent the flow of electrical current
   c) To transform voltage
   d) To store electrical energy

3. **Which semiconductor material is most commonly used in electronic devices?**

   a) Silicon
   b) Germanium
   c) Gallium arsenide
   d) Carbon

4. **What type of switch would you use to control a light from both ends of a hallway?**

   a) Single-pole switch
   b) Double-pole switch
   c) Three-way switches
   d) Four-way switch

5. **According to the NEC, GFCI protection is required in all of the following locations EXCEPT:**

   a) Bathrooms
   b) Outdoor receptacles
   c) Kitchen countertops
   d) Bedrooms

6. **What is the primary advantage of an LED dimmer over a traditional dimmer?**

   a) It costs less
   b) It works with all types of lighting
   c) It produces less heat
   d) It never requires a neutral connection

7. **What type of motor is most commonly used in residential ceiling fans?**

   a) Three-phase induction motor
   b) DC motor
   c) Single-phase capacitor-start motor
   d) Synchronous motor

8. **In a motor control circuit, what device protects against sustained overload conditions?**

   a) Circuit breaker
   b) Fuse
   c) Thermal overload relay
   d) Contactor

9. **Which of the following is NOT a typical component in a control circuit?**

   a) Limit switch
   b) Transformer
   c) Circuit breaker
   d) Push button

10. **What is the primary advantage of a solid-state relay over an electromagnetic relay?**

   a) Higher voltage capacity
   b) No moving parts to wear out
   c) Less expensive
   d) Simpler to install

11. **True or False: Aluminum conductors have higher conductivity than copper conductors of the same size.**

   a) True
   b) False

12. **Which of the following is a characteristic of semiconductors?**

   a) Their conductivity decreases as temperature increases
   b) They allow current flow in both directions equally
   c) Their conductivity can be precisely controlled by adding impurities
   d) They have lower resistance than copper

13. **What type of switch is required to control a light from three different locations?**

   a) One single-pole and two three-way switches
   b) Three three-way switches
   c) Two three-way switches and one four-way switch
   d) Three four-way switches

14. **What is the purpose of a tamper-resistant receptacle?**

    a) To prevent unauthorized removal of the receptacle
    b) To prevent insertion of foreign objects into the receptacle
    c) To protect against ground faults
    d) To protect against arc faults

15. **In a three-phase motor circuit, what happens if one phase is lost during operation?**

    a) The motor continues to run normally
    b) The motor immediately stops
    c) The motor continues to run but draws excessive current on the remaining phases
    d) The motor accelerates uncontrollably

16. **Which of the following sensors would be best for detecting the level of water in a tank?**

    a) Photoelectric sensor
    b) Float switch
    c) Proximity sensor
    d) Limit switch

17. **What does the term "normally open" (NO) mean when referring to a relay contact?**

    a) The contact is open when the relay is energized
    b) The contact is open when the relay is not energized
    c) The contact is closed during normal operation
    d) The contact opens only during fault conditions

18. **True or False: A GFCI device protects against arc faults that could cause fires.**

    a) True
    b) False

19. **What type of lighting control would be most appropriate for a storage closet where lights are frequently left on?**

    a) Dimmer switch
    b) Timer switch
    c) Occupancy sensor
    d) Daylight sensor

20. **Which of the following is NOT a type of AC motor?**

  a) Split-phase motor
  b) Shunt-wound motor
  c) Synchronous motor
  d) Squirrel-cage motor

Answer Key and Explanations

1. **Answer: c) Silver**

   While copper is more commonly used due to cost considerations, silver has the highest electrical conductivity of all metals (106% compared to copper's standard of 100%).

2. **Answer: b) To prevent the flow of electrical current**

   Insulators have tightly bound electrons that resist movement, preventing current flow and protecting against shock hazards and short circuits.

3. **Answer: a) Silicon**

   Silicon accounts for over 95% of semiconductor devices due to its abundance, stable properties, and suitable band gap for most applications.

4. **Answer: c) Three-way switches**

   Three-way switches are specifically designed to control a light from two different locations. Each three-way switch has three terminals: one common and two travelers.

5. **Answer: d) Bedrooms**

   The NEC requires GFCI protection in locations with increased shock risk, including bathrooms, kitchens (countertop receptacles), and outdoor locations, but not in bedrooms.

6. **Answer: c) It produces less heat**

   LED dimmers use solid-state electronics that generate less heat than traditional resistive dimmers, making them more efficient and safer.

7. **Answer: c) Single-phase capacitor-start motor**

   Most residential ceiling fans use single-phase capacitor-start motors because they're quiet, efficient, and can operate at different speeds.

8. **Answer: c) Thermal overload relay**

   Thermal overload relays are specifically designed to detect the sustained moderate overcurrent conditions typical of motor overloads, whereas circuit breakers and fuses primarily protect against short circuits.

9. **Answer: c) Circuit breaker**

   Circuit breakers are part of the power circuit, not the control circuit. They protect against overcurrent but don't typically participate in the control functions.

10. **Answer: b) No moving parts to wear out**

    Solid-state relays use semiconductor components instead of mechanical contacts, eliminating wear and extending operational life.

11. **Answer: b) False**

    Aluminum has approximately 61% of the conductivity of copper. This is why aluminum conductors must be larger than copper conductors for the same current-carrying capacity.

12. **Answer: c) Their conductivity can be precisely controlled by adding impurities**

    Through the process of doping, the electrical properties of semiconductors can be precisely controlled by adding specific impurities to create N-type or P-type materials.

13. **Answer: c) Two three-way switches and one four-way switch**

    To control a light from three locations, you need two three-way switches (at the beginning and end of the circuit) and one four-way switch between them.

14. **Answer: b) To prevent insertion of foreign objects into the receptacle**

    Tamper-resistant receptacles have internal shutters that block access to the contacts unless both slots are engaged simultaneously by a proper plug.

15. **Answer: c) The motor continues to run but draws excessive current on the remaining phases**

    This condition, known as "single-phasing," causes the motor to continue running on the remaining phases but with excessive current that can lead to overheating and damage.

16. **Answer: b) Float switch**

    Float switches are specifically designed to detect liquid levels and are commonly used in tanks and sumps.

17. **Answer: b) The contact is open when the relay is not energized**

    A normally open (NO) contact has an open circuit when the relay coil is not energized and closes when the coil is energized.

18. **Answer: b) False**

    GFCI devices protect against ground faults (current leakage to ground) that pose shock hazards. AFCI devices, not GFCIs, protect against arc faults that could cause fires.

19. **Answer: c) Occupancy sensor**

    Occupancy sensors automatically turn lights off when a space is vacant, making them ideal for areas where lights are frequently left on accidentally.

20. **Answer: b) Shunt-wound motor**

    Shunt-wound motors are DC motors, not AC motors. Split-phase, synchronous, and squirrel-cage are all types of AC motors.

Assessment

Count your correct answers:

- 18-20 correct: Excellent! You have a strong understanding of electrical components and devices.
- 15-17 correct: Good job! Review the questions you missed to strengthen your knowledge.
- 12-14 correct: You're on the right track, but should review the chapter more thoroughly.
- Below 12: Take time to re-read the chapter and work through the examples again.

Remember that understanding these components is essential for both your journeyman exam and your work in the field. These devices form the building blocks of all electrical systems you'll encounter throughout your career.

7

Advanced Electrical Concepts

The difference between an adequate electrician and an exceptional one isn't found in the basics—it's revealed in the mastery of advanced electrical principles. I've watched brilliant minds falter when confronted with power factor complications, while electricians with half their education but twice their practical wisdom solved problems that appeared unsolvable.

At a paper mill, a motor control system failed catastrophically despite perfect textbook calculations. The oversight? Power factor compensation. That single gap in understanding cost the company R4.3 million and three days of production. Theory without practical application is dangerous—impressive until something goes wrong.

The concepts in this chapter aren't academic exercises. They're solutions to problems that have humbled even veterans with decades of experience. Harmonics issues leave experienced professionals bewildered when they can't name what they're seeing, let alone fix it. Systems designed flawlessly on paper collapse when confronted with the harsh realities of transient loads and power quality fluctuations.

Every electrical mishap teaches something that textbooks can't convey. That capacitive discharge burn? A lesson no equation could properly explain. Those permanently damaged fingers? A reminder of what happens when fault current calculations go wrong. These aren't stories to frighten you—they emphasize that advanced concepts have concrete consequences, both harmful and beneficial.

As we explore power factor correction, harmonic mitigation, transient analysis, and complex impedance, remember that these aren't just formulas to memorize. They're essential tools that, when properly understood, become as vital to your work as your multimeter. They'll help you diagnose problems others miss entirely, prevent costly failures, and design systems with confidence.

7.1 Power Factor and Efficiency

"I don't understand it," the facility manager told me, pointing to his utility bill. "We installed all new equipment last year, but our electrical costs keep going up even though production hasn't increased."

When I checked the bill, the culprit was clear —a power factor penalty that was costing thousands of dollars monthly. The new equipment had significantly reduced the facility's power factor, something the manager had never considered when making purchase decisions.

This scenario plays out in facilities nationwide. While many electricians focus on meeting minimum code requirements for safety, truly skilled professionals understand that proper electrical design also addresses efficiency and power quality. Poor power factor and inefficient systems don't just waste energy—they increase operating costs, reduce equipment capacity, and can even lead to premature component failure.

In this section, we'll explore power factor correction, energy efficiency measures, and how these considerations impact both system design and operating costs. These concepts may not be the most exciting aspects of electrical work, but they often determine whether a client views your electrical installation as a success or a costly mistake.

7.2 Power Factor

Power factor is a measure of how effectively electrical power is being used. It is the ratio of real power (kW) to apparent power (kVA) in an electrical system.

Most electricians understand voltage, current, and resistance, but power factor introduces a more complex dimension to electrical systems. It forces us to recognize that not all power delivered to a facility performs useful work. This concept becomes particularly important in commercial and industrial installations where power factor penalties can significantly impact operating costs.

Understanding Power Factor

Power Factor (PF) = Real Power (kW) / Apparent Power (kVA)

To grasp power factor, we need to understand three types of power:

- **Real Power (kW)** performs actual work —running motors, creating light, or generating heat. This is the power component that accomplishes useful tasks and is measured in kilowatts (kW).

- **Reactive Power (kVAR)** doesn't perform work but instead creates and sustains electromagnetic fields in inductive equipment like motors and transformers. Reactive power oscillates between the source and the load without being consumed. It's measured in kilovolt-amperes reactive (kVAR).
- **Apparent Power (kVA)** is the vector sum of real and reactive power, representing the total power supplied by the utility. This is what the electrical system must be sized to handle, regardless of how much real work is being done. It's measured in kilovolt-amperes (kVA).

When reactive power is present in a system, the current waveform shifts out of phase with the voltage waveform. In inductive loads, current lags behind voltage, while in capacitive loads, current leads voltage. This phase difference is what creates power factor issues.

Importance of Power Factor

A low power factor indicates poor utilization of electricity, leading to wasted energy and higher utility bills. A power factor of 1.0 is ideal, meaning all the power supplied is being used efficiently.

The consequences of low power factor include:

- **Higher Utility Bills**: Many utilities impose penalties on facilities with power factors below a specified threshold (typically 0.85 or 0.90). These penalties can add 10-20% to monthly electric bills.
- **Reduced System Capacity**: Low power factor means the electrical system carries more current than necessary for the same amount of work. This reduces the available capacity of transformers, conductors, and other distribution equipment.
- **Increased Voltage Drop**: Higher currents associated with low power factor create greater voltage drops in distribution systems, potentially causing equipment performance issues.
- **Higher Heat Losses**: Increased current flow generates more heat in conductors and equipment, reducing efficiency and potentially shortening equipment life.

Power Factor Correction

Devices like capacitors and synchronous condensers are used to improve the power factor by compensating for the lag caused by inductive loads.

Correction methods include:

- **Capacitor Banks**: The most common solution, capacitors provide leading reactive power to offset the lagging reactive power of inductive loads. These can be installed on individual equipment, distribution panels, or at the service entrance.

- **Synchronous Condensers**: Special motors that can be adjusted to produce reactive power for correction purposes.
- **Active Power Factor Correction**: Electronic systems that dynamically adjust correction based on changing loads.
- **Variable Frequency Drives with Built-in Correction**: Modern VFDs often include power factor correction capabilities.

When implementing power factor correction, placement matters. Correction at the equipment level reduces reactive power flow throughout the distribution system, while service entrance correction only addresses utility penalties but doesn't reduce internal system losses.

For proper sizing of correction equipment, measurements should be taken under typical operating conditions to determine the actual power factor and correction requirements. The goal is typically to achieve a power factor of 0.95 or better to avoid utility penalties and optimize system efficiency.

7.3 Efficiency

Efficiency in electrical systems refers to the ratio of useful power output to the total power input, expressed as a percentage.

Every electrical device and system converts energy from one form to another, but none does so perfectly. Some energy is always lost in the process—typically as heat, vibration, or noise. Understanding efficiency helps electricians design systems that minimize these losses, saving energy and reducing operating costs.

Measuring Efficiency

Efficiency (%) = (Useful Power Output / Total Power Input) × 100

This straightforward formula reveals the percentage of input power that's successfully converted to useful output. The remaining energy becomes waste that doesn't contribute to the device's intended purpose. For example:

In a motor that consumes 10 kW of electrical power but produces only 8.5 kW of mechanical power at its shaft, the efficiency is: (8.5 kW / 10 kW) × 100 = 85%

This means 15% of the input power becomes waste—mostly heat in the motor windings and core, friction in bearings, and windage losses.

Different equipment types have different efficiency ranges. Modern premium-efficiency motors might reach 95% efficiency, while transformers can exceed 98%. Even LED lighting, considered highly efficient, typically converts only 40-50% of input power to visible light, with the remainder becoming heat.

Importance of Efficiency

High efficiency means less energy is wasted in the system, reducing operational costs and improving performance.

The benefits of high-efficiency systems include:

- **Reduced Operating Costs**: Energy savings translate directly to lower utility bills. For equipment that operates continuously, even small efficiency improvements can yield substantial cost savings over time.
- **Decreased Heat Generation**: Higher efficiency means less energy converted to waste heat, reducing cooling requirements in facilities and extending equipment life by operating at lower temperatures.
- **Smaller Power Distribution Systems**: More efficient equipment draws less current for the same output, potentially allowing for smaller conductors, transformers, and distribution equipment.
- **Environmental Benefits**: Lower energy consumption reduces the environmental impact of electrical systems, including reduced carbon emissions from power generation.
- **Compliance with Regulations**: Energy codes and standards increasingly mandate minimum efficiency levels for electrical equipment and systems.

Factors Affecting Efficiency

Resistance in conductors, losses in transformers, and motor inefficiency can all contribute to reduced system efficiency.

Major sources of losses include:

- **Conductor Resistance**: As current flows through conductors, energy is converted to heat through I^2R losses. These losses increase with the square of the current, so doubling current quadruples resistance losses.
- **Transformer Losses**: These include core losses (present whenever the transformer is energized) and copper losses (proportional to the square of the load current).
- **Motor Losses**: These include stator and rotor copper losses, core losses, friction and windage losses, and stray load losses. Motor efficiency typically peaks at 75-100% of rated load and

decreases significantly at lighter loads.

- **Conversion Losses**: Each energy conversion step introduces inefficiencies. For example, in an LED lighting system, losses occur in the driver converting AC to DC and in the LED converting electricity to light.
- **Distribution System Losses**: Voltage drops in feeders and branch circuits represent energy converted to heat rather than delivered to loads.

7.4 AC vs. DC Systems and Three-Phase Power

The debate between AC and DC power systems dates back to the late 1800s, when Edison championed direct current while Tesla and Westinghouse advocated for alternating current. Though that historic "War of Currents" was ultimately won by AC power for most applications, both systems remain vital in modern electrical work.

Today's electrical landscape is more complex than ever. While utility distribution remains firmly in AC territory, many modern devices internally convert power to DC. From the vehicle charging station outside to the server racks running cloud computing, DC power is experiencing a renaissance. Meanwhile, three-phase power continues to drive industrial equipment, offering advantages in efficiency and power density that single-phase systems can't match.

This section explores the key differences between AC and DC systems, the advantages and applications of each, and the fundamentals of three-phase power that make it the backbone of industrial electrical systems worldwide.

7.5 AC vs. DC Systems

Alternating Current (AC) systems are the most common form of electrical power distribution in residential, commercial, and industrial settings. The direction of current flow alternates periodically.

AC power became dominant largely because of its transmission advantages. In an AC system, current and voltage oscillate in a sinusoidal pattern, typically completing 60 cycles per second (60 Hz) in North America or 50 Hz in many other countries. This oscillation creates unique properties that shaped our entire electrical infrastructure.

Advantages of AC Systems

AC is more efficient for long-distance transmission due to lower power losses. It's easy to step up or step down voltage using transformers.

The ability to use transformers represents AC's greatest advantage. Power transmission losses are proportional to current squared (I^2R), so transmitting power at high voltage and low current dramatically reduces these losses. AC allows for simple and efficient voltage transformation without moving parts:

- Power plants generate at moderate voltages (typically 13.8-24kV).
- Step-up transformers increase this to transmission voltages (138kV-765kV).
- Substations use step-down transformers to reach distribution voltages (4kV-35kV).
- Local transformers further reduce voltage to utilization levels (120V-480V).

This voltage flexibility made large-scale power distribution economically viable and remains AC's primary advantage over DC systems.

Applications of AC Power

AC is used for powering homes, businesses, and large electrical grids.

The vast majority of electrical devices and systems you'll work with operate on AC power:

- Residential and commercial buildings receive AC power from utilities
- Industrial motors and equipment typically run on AC power
- Lighting systems traditionally operate on AC
- Heating elements work equally well with AC or DC

Motors represent a particularly important AC application. Induction motors, which dominate industrial applications, operate directly from AC power without requiring the brushes or commutators needed in traditional DC motors.

Disadvantages of AC Systems

AC can create electromagnetic interference (EMI) and is less efficient for certain types of electronics.

The constantly changing direction of current in AC systems creates oscillating electromagnetic fields that can induce currents in nearby conductors. This can cause:

- Interference with sensitive electronic equipment.
- Increased proximity effect in conductors.
- Skin effect, where current tends to flow on conductor surfaces at higher frequencies.
- Reactive power issues with inductive and capacitive loads.

Modern electronics predominantly operate on DC power internally. Devices like computers, televisions, and LED lighting contain power supplies that convert AC to DC, introducing conversion losses. These losses become significant in data centers and other facilities with high concentrations of electronic equipment.

Despite these disadvantages, AC remains the standard for power distribution due to its transmission advantages and the massive infrastructure built around it. Understanding both the strengths and limitations of AC systems helps electricians make appropriate design and installation decisions for different applications.

Direct Current (DC)

In DC systems, the current flows in one direction.

Direct current represents the simplest form of electricity—electrons move consistently from the negative terminal to the positive terminal. Unlike alternating current, which changes direction many times per second, DC maintains a steady, unchanging flow. This is the type of current produced by batteries, solar cells, and fuel cells, where chemical reactions create a consistent push of electrons in one direction.

Visualize water flowing through a pipe in only one direction—this analogy helps understand DC. The pressure (voltage) may vary, but the flow direction remains constant. This consistent directional flow gives DC power distinct characteristics that make it essential for many applications.

Advantages of DC Systems

DC provides stable voltage and is ideal for powering electronics like computers, batteries, and electric vehicles.

- The non-oscillating nature of DC power offers several key benefits:
- Electronic components like microprocessors require steady voltage levels that DC

naturally provides.
- Battery systems inherently produce and store energy as DC, creating natural compatibility.
- DC eliminates power quality issues related to frequency and phase relationships.
- Motor control can be more precise with DC, allowing exact speed regulation.

Disadvantages of DC Systems

DC is less efficient for long-distance transmission and requires more complex voltage conversion methods.

The primary limitations that restricted DC adoption historically include:

Transforming DC voltage levels was historically difficult. While AC can be easily stepped up or down using transformers, DC requires motor-generator sets for voltage conversion until modern power electronics emerged.

Circuit interruption is more challenging with DC. When interrupting AC, the natural zero-crossing points of the sine wave help extinguish the arc that forms between contacts. DC lacks these zero crossings, requiring larger, more sophisticated circuit breakers.

While high-voltage DC transmission now exists for specific applications like underwater cables and very long distances, it requires expensive converter stations at both ends—a sharp contrast to the relatively simple transformers used in AC systems.

As an electrician, you'll increasingly work with systems that combine both AC and DC elements, particularly in renewable energy installations, electric vehicle charging, and modern building systems that aim to reduce energy conversion losses.

7.5 Three-Phase Power

Three-phase power is a type of AC power generation and distribution that uses three conductors, each carrying an alternating current that is offset by one-third of a cycle.

When you examine a three-phase system, you'll find three separate AC waveforms that are 120 electrical degrees apart from each other. This staggered relationship between phases creates unique properties that make three-phase power indispensable for industrial and commercial applications. While single-phase power delivers energy in pulses that drop to zero 120 times per second (at 60Hz), three-phase power delivers a constant, steady stream of power that never drops to zero.

This fundamental difference explains why substantial loads like large motors, industrial equipment, and commercial HVAC systems overwhelmingly use three-phase power.

Advantages of Three-Phase Power

Three-phase power is more efficient than single-phase power, providing a constant flow of power, reducing voltage dips, and ensuring smoother operation of motors and equipment.

The efficiency advantages of three-phase systems are substantial:

- Three-phase motors produce a rotating magnetic field naturally, eliminating the need for starting capacitors or other special starting arrangements.
- Power delivery is more constant, with 1.732 (√3) times more power-carrying capacity per conductor compared to single-phase systems.
- Three-phase equipment typically runs more smoothly, with less vibration and noise.
- Voltage imbalances under varying loads tend to be less problematic.

Perhaps most impressively, a three-phase system can deliver 1.73 times more power than a single-phase system with the same conductor size. This means three-phase distribution requires less copper or aluminum to deliver the same amount of power, a significant economic advantage.

Applications of Three-Phase Power

Three-phase power is widely used in industrial and commercial settings, particularly where large motors and heavy equipment are in operation.

You'll encounter three-phase systems in:

Industrial facilities where large motors drive manufacturing equipment, pumps, compressors, and material handling systems. These environments benefit from the higher efficiency and smoother operation of three-phase motors, particularly under heavy or varying loads.

Commercial buildings with substantial HVAC equipment, elevators, and large lighting loads. Modern commercial buildings typically receive three-phase service even if many internal loads are single-phase.

Data centers, where the high power density and continuous operation make the efficiency advantages of three-phase distribution particularly valuable. Even small efficiency improvements translate to significant cost savings in these energy-intensive facilities.

Distribution systems, where three-phase power is used throughout the grid, from generation to high-voltage transmission to local distribution. Even residential customers receive power from a three-phase distribution system, though it's converted to single-phase at the service drop.

7.6 Harmonics, Grounding, and Transient Voltage

The hospital's critical care wing had experienced mysterious equipment failures for months. Electronic monitors would reset without warning. Diagnostic equipment produced inconsistent readings. Despite replacing numerous devices, the problems persisted until an electrician discovered the hidden culprit: harmonic distortion from nearby variable frequency drives was corrupting the clean power these sensitive instruments required.

Power quality issues like harmonics, improper grounding, and transient voltages rarely announce themselves with obvious symptoms like outages or tripped breakers. Instead, they lurk in electrical systems, causing subtle, intermittent problems that frustrate troubleshooting efforts. These invisible power quality problems can degrade equipment performance, reduce efficiency, and sometimes cause catastrophic failures—all while standard voltage and current measurements appear normal.

As buildings incorporate more non-linear loads and sensitive electronic equipment, these once-obscure electrical phenomena have become central concerns for modern electricians. Understanding these complex power quality issues—how to detect them, prevent them, and mitigate their effects—has become an essential skill set that distinguishes exceptional electrical professionals from the rest.

7.7 Harmonics

Harmonics are voltage or current waveforms that operate at frequencies that are multiples of the fundamental frequency (usually 60Hz for AC systems in the U.S.).

In an ideal world, AC power would follow a perfect sine wave pattern. But in real electrical systems, particularly those with modern electronic equipment, the actual waveform becomes distorted. These distortions create additional frequency components—harmonics—that ride on top of the fundamental frequency.

Think of musical overtones, when you hear a guitar string, you're not just hearing the fundamental note but also higher-frequency overtones that give the instrument its distinctive sound. Similarly, electrical harmonics are higher-frequency components that distort the pure sine wave of AC power.

Each harmonic is identified by its order number, which indicates its frequency relationship to the fundamental:

- 2nd harmonic = 120Hz (60Hz × 2)
- 3rd harmonic = 180Hz (60Hz × 3)
- 5th harmonic = 300Hz (60Hz × 5)
- 7th harmonic = 420Hz (60Hz × 7)

Different harmonic orders create different effects in electrical systems. Odd-order harmonics (3rd, 5th, 7th, etc.) are typically more prevalent than even-order harmonics in balanced electrical systems.

Causes of Harmonics

Non-linear loads such as computers, variable frequency drives (VFDs), and certain types of lighting cause harmonics in electrical systems.

A non-linear load draws current in pulses rather than in a smooth sinusoidal pattern. These pulse-drawing devices include:

Electronic power supplies found in computers, servers, and most modern office equipment. These typically use switch-mode power supplies that draw current in short pulses during each cycle, creating significant 3rd, 5th, and 7th harmonics.

Variable frequency drives used for motor control. VFDs are major sources of harmonics, particularly 5th and 7th order, due to the rectification process that converts AC to DC before creating variable-frequency output.

LED and fluorescent lighting with electronic ballasts. Though individual fixtures may produce relatively small harmonic distortion, the cumulative effect of hundreds of fixtures can be substantial.

Uninterruptible power supplies (UPS), especially those without power factor correction, can generate significant harmonics during both charging and discharging cycles.

Arc furnaces and welding equipment, which create variable and unpredictable current demands that generate broadband harmonic content.

Effects of Harmonics

Harmonics can lead to overheating of electrical equipment, increased energy consumption, and equipment malfunctions.

The presence of harmonics creates several problems in electrical systems:

Conductor and transformer overheating occurs because harmonic currents increase the RMS current without contributing useful power. The skin effect, where current tends to flow on conductor surfaces at higher frequencies, further exacerbates heating issues with higher-order harmonics.

Neutral conductor overloading is particularly problematic with 3rd harmonics. In a balanced three-phase system, fundamental currents cancel in the neutral. However, 3rd harmonics (and other triplen harmonics—3rd, 9th, 15th, etc.) add together in the neutral, potentially causing dangerous overheating in systems where the neutral is sized the same as phase conductors.

Motor efficiency reduction and overheating result from harmonic voltages creating opposing magnetic fields in the motor, increasing losses, and reducing torque.

Nuisance tripping of circuit breakers occurs when harmonics cause peak currents to exceed breaker trip settings even though the fundamental current is within normal ranges.

Interference with electronic equipment can cause control system malfunctions, communications errors, and flickering displays. This is particularly problematic in medical facilities, laboratories, and data centers.

Power factor correction capacitors may form resonant circuits with system inductance, amplifying specific harmonic frequencies and potentially leading to capacitor failure.

Mitigating Harmonics

Using filters, power conditioners, and equipment designed to minimize harmonic distortion can reduce the negative effects.

Several strategies help manage harmonic issues:

Passive harmonic filters use inductors and capacitors tuned to specific harmonic frequencies, providing a low-impedance path that diverts harmonics away from sensitive equipment.

Active harmonic filters electronically monitor distortion and inject counteracting currents to cancel harmful harmonics, offering more dynamic and comprehensive harmonic mitigation.

Line reactors (inductors) installed with VFDs add impedance that smooths current draw and reduces harmonic generation at the source.

K-rated transformers are specifically designed with additional capacity to handle harmonic loads without overheating. The K-factor rating indicates the transformer's ability to handle harmonic content.

Phase-shifting transformers use multiple windings with different phase relationships to cancel certain harmonics. Delta-wye transformers naturally trap 3rd harmonics in the delta winding.

Oversized neutral conductors (typically doubled to 200% in areas with significant harmonics) prevent neutral overheating from triplen harmonics.

Modern equipment with built-in harmonic mitigation, such as VFDs with DC or AC line reactors and active front-end technology, can significantly reduce harmonic generation at the source.

The IEEE 519 standard provides recommended practices and requirements for harmonic control, establishing limits for both individual harmonic components and total harmonic distortion (THD) at the point of common coupling with the utility.

As electrical systems incorporate more non-linear loads, harmonic management becomes increasingly important. Proper measurement, analysis, and mitigation strategies help ensure system reliability, efficiency, and compliance with industry standards.

Transient Voltage

Transient voltage is a temporary, high-voltage spike that can damage electrical equipment.

These short-duration surges in voltage occur in milliseconds or even microseconds, but their effects can be devastating. Despite their brief nature, transients can reach several thousand volts, overwhelming standard insulation and protective devices designed for steady-state conditions. Understanding these events helps electricians implement appropriate protection strategies.

Transient voltages fall into several categories based on their magnitude, duration, and frequency content:

- Impulse transients: Very fast spikes with rise times in microseconds.
- Oscillatory transients: Voltage spikes that swing between positive and negative values
- Voltage swells: Temporary increases in RMS voltage lasting several cycles

Causes of Transients

Lightning strikes, switching of electrical loads, and equipment failures can all generate transient voltage.

Natural events represent the most dramatic source of transients. A direct lightning strike can introduce hundreds of thousands of volts into electrical systems, but even distant strikes can induce significant transients through electromagnetic coupling. These events typically create the most severe transients that electrical systems encounter.

Internal switching operations are more common causes of transients:

- Motor starts and stops create sudden changes in current flow
- Capacitor bank switching in utility systems generates oscillatory transients
- Circuit breaker and switch operations produce arcing that generates high-frequency transients
- Transformer energizing creates inrush currents that distort voltage waveforms

Equipment failures also generate transient events:

- Insulation breakdown leading to arcing
- Short circuits that collapse voltage before protective devices operate
- Ground faults that shift neutral-to-ground relationships
- Power factor correction equipment failures

Effects of Transient Voltage

Transients can cause equipment damage, data loss, and reduced equipment lifespan.

The impact of transients varies based on both the transient characteristics and the equipment's vulnerability:

Immediate equipment failure occurs when transient voltage exceeds the insulation breakdown strength of components. Sensitive semiconductor devices can fail at voltage levels just twice their rated value, while insulation in motors or transformers might withstand significantly higher overvoltages.

Cumulative damage often goes unnoticed but progressively degrades equipment. Repeated lower-level transients cause insulation stress that eventually leads to failure, similar to bending a paper clip repeatedly until it breaks. This makes transients particularly insidious—the damage accumulates silently until catastrophic failure occurs.

Data corruption in electronic systems happens when transients affect signal levels in digital circuits. A voltage spike that doesn't physically damage hardware might still corrupt data being processed or transmitted, causing system crashes or errors.

Process disruption occurs when transients cause momentary malfunctions in control systems. Even brief disruptions to manufacturing equipment, medical devices, or safety systems can have serious consequences.

Protection Against Transient Voltage

Surge protectors, transient voltage suppressors (TVS), and proper grounding can help protect systems from transient voltage damage.

A layered approach provides the most effective protection:

Service entrance protection serves as the first line of defense, typically using surge arresters that can handle the high energies associated with lightning and utility switching events. These devices divert high-energy surges to ground before they enter the building's electrical system.

Distribution level protection is installed at panelboards and subpanels, typically using TVSS (Transient Voltage Surge Suppression) devices. These systems handle moderate energy surges that pass through or originate beyond the service entrance protection.

Point-of-use protection includes plug-in surge protectors and built-in protection circuits in sensitive equipment. These devices handle lower-energy transients that originate within the building or pass through upstream protective devices.

Proper system grounding is essential for all surge protection methods. Without a low-impedance path to ground, even the best surge protective devices cannot effectively divert transient energy away from sensitive equipment.

Zone-based protection strategies create progressive levels of protection as power enters and is distributed throughout a facility. Each zone boundary includes surge protection appropriate to that level, with coordination between stages to ensure proper energy handling.

For critical facilities like data centers, hospitals, and industrial control systems, transient voltage protection becomes a crucial design consideration. The cost of comprehensive protection is minimal compared to the potential losses from equipment damage, downtime, and data loss that transient voltage events can cause.

7.8 Quiz: Testing Your Knowledge of Advanced Electrical Concepts

The most dangerous electrical concepts are often the ones you can't see. Power factor issues silently increase utility bills, harmonic distortion gradually degrades equipment, and transient voltage problems lurk undetected until sudden failures occur. Let's test your grasp of these invisible but crucial electrical phenomena.

This quiz goes beyond basic theory to challenge your practical understanding of these advanced concepts. Can you identify power quality issues from their symptoms? Do you know which mitigation strategies work for different problems? The questions below will test not just what you've memorized, but how well you can apply these concepts to real-world situations—exactly what you'll need to do as a journeyman electrician.

Put your knowledge to the test with these scenarios and questions that mirror the complex electrical challenges you'll face in the field.

Practice Questions

1. **A manufacturing facility has an apparent power (S) of 500 kVA and a real power (P) of 425 kW. What is the power factor?**

   a) 0.85
   b) 0.78
   c) 0.92
   d) 1.18

2. **Which of the following loads typically causes a lagging power factor?**

   a) Electric motors
   b) Capacitor banks
   c) LED lighting with power factor correction
   d) Synchronous motors overexcited

3. **What is the primary advantage of three-phase power over single-phase power?**

   a) Lower voltage
   b) Higher voltage
   c) More constant power delivery
   d) Simpler wiring

4. **A 10 HP three-phase motor has an efficiency of 89%. If the motor is fully loaded, approximately how much power does it convert to heat rather than mechanical work?**

   a) 0.89 kW
   b) 1.0 kW
   c) 0.92 kW
   d) 0.93 kW

5. **True or False: A power factor of 0.75 leading would typically require the addition of inductors for correction.**

   a) True
   b) False

6. **Which harmonic order is particularly problematic for neutral conductors in a balanced three-phase system?**

   a) 2nd harmonic
   b) 3rd harmonic
   c) 5th harmonic
   d) 7th harmonic

7. **Which device is most likely to generate significant harmonic currents?**

   a) Incandescent lighting
   b) Resistive heater
   c) Variable frequency drive
   d) Direct-on-line motor starter

8. **What is the primary advantage of DC power over AC power?**

   a) Easier voltage transformation
   b) More efficient for long-distance transmission
   c) Direct compatibility with batteries and electronic devices
   d) Simpler circuit protection

9. **A three-phase circuit carries 30 amperes per phase at 208V. What is the total power if the power factor is 0.9?**

   a) 5.6 kW
   b) 9.7 kW
   c) 5.2 kW
   d) 8.7 kW

10. **Which of the following would NOT be considered a source of transient voltage?**

    a) Lightning strike
    b) Motor startup
    c) Capacitor bank switching
    d) Voltage drop due to distance

11. **What is the main function of a surge protector in an electrical system?**

    a) To provide backup power during outages
    b) To regulate voltage during normal operation
    c) To divert transient voltage surges away from sensitive equipment
    d) To correct power factor issues

12. **True or False: The harmonics generated by non-linear loads always result in a higher RMS current than would be expected from the fundamental frequency alone.**

    a) True
    b) False

13. **An electrical system has a power factor of 0.75. What percentage of the apparent power is doing useful work?**

    a) 25%
    b) 75%
    c) 56%
    d) 33%

14. **Which of the following is NOT a typical effect of poor power factor?**

    a) Increased utility bills
    b) Reduced system capacity
    c) Improved voltage regulation
    d) Higher I^2R losses

15. **In a wye-connected three-phase system, what is the relationship between phase voltage and line voltage?**

    a) Phase voltage = Line voltage
    b) Phase voltage = Line voltage × 1.732
    c) Phase voltage = Line voltage ÷ 1.732
    d) Phase voltage = Line voltage × 3

Answer Key and Explanations

1. **Answer: a) 0.85**

   Power Factor = Real Power ÷ Apparent Power = 425 kW ÷ 500 kVA = 0.85. This indicates that 85% of the apparent power is being converted to useful work.

2. **Answer: a) Electric motors**

   Electric motors create a lagging power factor because they are inductive loads. The current lags behind the voltage due to the magnetic fields that must be established before energy is converted to mechanical work.

3. **Answer: c) More constant power delivery**

   Three-phase power provides a constant power delivery that never drops to zero, unlike single-phase power, which pulsates. This creates smoother operation for motors and more consistent power for other loads.

4. **Answer: c) 0.92 kW**

   A 10 HP motor converts approximately 7.46 kW of electrical power to mechanical power when fully loaded. With 89% efficiency, it draws $7.46 \div 0.89 = 8.38$ kW from the electrical system. The difference ($8.38 - 7.46 = 0.92$ kW) is lost as heat.

5. **Answer: b) False**

   A leading power factor indicates an excess of capacitive reactance in the system. To correct a leading power factor, you would add inductors (reactors), not capacitors, which would make the problem worse.

6. **Answer: b) 3rd harmonic**

   Third harmonics (and other triplen harmonics 3rd, 9th, 15th, etc.) add together in the neutral of a three-phase four-wire system instead of canceling as the fundamental frequencies do. This can cause neutral conductors to carry more current than the phase conductors.

7. **Answer: c) Variable frequency drive**

   VFDs convert AC to DC and then back to variable-frequency AC using power electronics that draw current in non-sinusoidal patterns, generating significant harmonics.

8. **Answer: c) Direct compatibility with batteries and electronic devices**

   DC power offers direct compatibility with batteries for energy storage and with the internal circuitry of electronic devices, which all operate on DC internally.

9. **Answer: b) 9.7 kW**

   Three-phase power = $\sqrt{3} \times V \times I \times PF = 1.732 \times 208V \times 30A \times 0.9 = 9{,}720$ watts or 9.7 kW.

10. **Answer: d) Voltage drop due to distance**

    Voltage drop due to distance is a steady-state condition resulting from conductor impedance, not a transient event. The other options all create sudden, temporary changes in voltage.

11. **Answer: c) To divert transient voltage surges away from sensitive equipment**

    Surge protectors are designed to detect and divert voltage spikes that exceed normal operating levels, protecting connected equipment from damage.

12. **Answer: a) True**

    Harmonics always increase the RMS current above what would be expected from the fundamental frequency alone. This is why equipment and conductors carrying harmonic currents must be derated.

13. **Answer: b) 75%**

    Power factor directly indicates the percentage of apparent power that is converted to real power. With a PF of 0.75, 75% of the apparent power is doing useful work.

14. **Answer: c) Improved voltage regulation**

    Poor power factor typically worsens voltage regulation due to increased current flow and corresponding voltage drops. The other options are all genuine consequences of poor power factor.

15. **Answer: c) Phase voltage = Line voltage ÷ 1.732**

    In a wye-connected system, the line voltage is √3 (approximately 1.732) times higher than the phase voltage. Therefore, phase voltage equals line voltage divided by 1.732.

Assessment

Count your correct answers:

- 13-15 correct: Excellent! You have a strong understanding of advanced electrical concepts.
- 10-12 correct: Good job! Review the questions you missed to strengthen your knowledge.
- 7-9 correct: You're on the right track, but should review the chapter more thoroughly.
- Below 7: Take time to re-read the chapter and work through the examples again.

These advanced electrical concepts are crucial for troubleshooting power quality issues and designing efficient electrical systems. Understanding these principles will help you excel both on the journeyman exam and in real-world applications.

8

Wiring and Circuitry Applications

Ask ten master electricians how to wire a three-way switch system, and you'll likely get eleven different answers. That's because wiring isn't just about connecting point A to point B—it's about crafting solutions that balance code requirements, physical constraints, future serviceability, and the realities of time and budget.

During my first solo commercial job, I watched in horror as my carefully planned conduit layout met its first real-world obstacle—a massive HVAC duct that wasn't on any blueprint. That day taught me that book knowledge only gets you halfway there. The real art of electrical work emerges when theory meets the unpredictable world of actual buildings, stubborn existing conditions, and the inevitable Murphy's Law of construction.

The difference between an ordinary electrician and an exceptional one often lies not in what they know about the code, but in how they apply that knowledge. Anyone can install a receptacle, but creating a thoughtfully designed kitchen circuit layout that anticipates future needs while minimizing voltage drop requires both experience and foresight.

This chapter focuses on the practical applications that transform theoretical electrical knowledge into functional systems. We'll examine proven approaches to common wiring scenarios, explore the logic behind effective circuit design, and share field-tested techniques that make installations both compliant and practical.

8.1 Branch Circuits and Feeders

The project manager handed me a set of drawings for a new medical office building and pointed to a cramped electrical room. "You've got 37 panels to fit in here," he said. "Good luck."

As I surveyed the space, I knew the real challenge wouldn't be mounting the panels—it would be designing a feeder and branch circuit system that balanced code requirements, future expansion needs, and the practical realities of installation.

Branch circuits and feeders are a part of the circulatory system of any electrical installation. Like arteries and capillaries in the human body, they distribute power from their source to the furthest reaches of a building. The main service feeders carry substantial current to distribution panels, which branch out to smaller feeders, which in turn supply branch circuits that power individual loads.

The distinctions between these circuit types—and knowing how to properly size, protect, and install them—are fundamental to electrical work. A well-designed system provides adequate capacity without waste, maintains proper voltage at all devices, and accommodates future growth without requiring major renovations. A poorly designed system, on the other hand, becomes a recurring source of problems: nuisance tripping, voltage drop issues, and premature equipment failure.

As we explore branch circuits and feeders, we'll examine not just the code requirements that govern them, but the practical considerations that make them work in real-world applications. The difference between theory and practice often becomes apparent in how we arrange, route, and protect these vital power distribution pathways.

8.2 Branch Circuits

A branch circuit is a portion of an electrical circuit that extends from the final overcurrent protection device (usually a circuit breaker) to the electrical devices or outlets in the system.

Branch circuits represent the final leg in an electrical distribution system, connecting the overcurrent protection device directly to the loads it serves. These circuits handle the day-to-day power delivery that makes buildings functional, from providing light in a hallway to powering critical equipment in an operating room. Their proper design, installation, and protection form the foundation of safe, reliable electrical systems.

Purpose of Branch Circuits

Branch circuits supply electrical power to specific loads like lights, outlets, and appliances.

The primary goal of branch circuit design is to deliver adequate power to each load while maintaining both safety and efficiency. Branch circuits must:

- Carry sufficient current without overheating
- Maintain voltage within acceptable limits for connected equipment
- Provide appropriate overcurrent protection
- Ensure proper grounding for safety
- Accommodate the specific needs of the loads they serve

Each branch circuit has a specific purpose within the overall electrical system. A well-designed system separates loads logically, allowing for maintenance without total power loss and providing appropriate capacity for each type of load.

8.3 Types of Branch Circuits

Lighting Circuits

Lighting circuits supply power to lighting fixtures and outlets for lamps. These circuits typically operate at 120V in residential applications and either 120V or 277V in commercial buildings.

The NEC requires general lighting calculations based on square footage:

- Dwelling units: 3 VA per square foot (NEC 220.12)
- Commercial spaces: Varies by occupancy type, ranging from 0.5 VA to 3 VA per square foot

Lighting circuits often serve multiple fixtures controlled by switches or lighting control systems. Modern lighting circuits must account for energy code requirements that may include occupancy sensing, daylight harvesting, or time-based controls.

Appliance Circuits

Appliance circuits are designed for power-hungry devices like dryers, stoves, and refrigerators.

The NEC requires specific provisions for these higher-demand loads:

- Small appliance circuits: Minimum of two 20A, 120V circuits for kitchen countertop receptacles.
- Laundry circuits: Dedicated 20A, 120V circuit for laundry equipment.
- Large appliances: Sized according to nameplate ratings plus any applicable demand factors.

Voltage requirements vary by appliance type:

- Most kitchen countertop appliances: 120V
- Electric ranges, ovens: 240V
- Clothes dryers: 120/240V

Dedicated Circuits

Dedicated circuits serve a single load, typically a high-power device such as an air conditioner or electric oven.

The NEC requires dedicated circuits for specific loads:

- Electric water heaters
- Electric space heating
- Dishwashers and garbage disposals (in many jurisdictions)
- HVAC equipment
- Electric vehicle charging equipment

Dedicated circuits prevent the problems that occur when high-demand loads share circuits with other equipment. They eliminate nuisance tripping when multiple devices start simultaneously and ensure voltage stability for sensitive equipment.

The proper selection, sizing, and protection of branch circuits directly impact both safety and performance. A circuit that's marginally sized might work initially but lead to voltage drop issues, overheating, or nuisance tripping as it ages or when loads increase. Conversely, grossly oversized circuits waste materials and add unnecessary cost to projects.

As you prepare for your journeyman exam, focus on the NEC requirements for minimum branch circuit sizing and the specific provisions for different load types and occupancies. The code establishes minimum requirements, but practical field experience often leads electricians to exceed these minimums in critical applications.

8.4 Feeders

A feeder is a set of conductors used to deliver power from the service equipment or distribution panel to a subpanel, distribution box, or load center.

While branch circuits connect directly to utilization equipment, feeders distribute power to downstream distribution points. They are the main highways of an electrical system, carrying substantial current before it branches onto the smaller local roads. These circuits handle the bulk power distribution within a building or campus, requiring careful planning to ensure reliability, efficiency, and code compliance.

Purpose of Feeders

Feeders supply electrical power from the main panel to secondary distribution points or loads.

The primary function of feeders is to efficiently distribute power throughout larger structures where a single panel would be impractical. A well-designed feeder system:

- Strategically places distribution points near load centers.
- Minimizes voltage drop across long distances.
- Accommodates future expansion needs.
- Maintains system reliability through appropriate redundancy.
- Allows for isolation of building sections for maintenance or emergencies.

In larger buildings, feeders create a hierarchical distribution system. Main feeders might supply motor control centers in an industrial facility, while smaller feeders deliver power to lighting panels distributed throughout the building. This tiered approach allows for more efficient power distribution and simplified circuit organization.

Feeder Design Considerations

Factors like conductor size, voltage drop, and overcurrent protection are key to proper feeder design.

Proper feeder sizing involves more than just matching current capacity to the load:

- Voltage drop becomes critical for long feeder runs, with the NEC recommending (not requiring) a maximum 3% drop for feeders alone.
- Ambient temperature, conduit fill, and installation method all affect allowable ampacity.
- Available fault current must be considered when selecting overcurrent protection.
- Physical routing challenges often constrain conductor size and raceway options.

The NEC addresses feeder calculations primarily in Article 220 Part III, with important provisions for demand factors that prevent unnecessary oversizing. For example, continuous lighting loads must be calculated at 125% of the actual load, while motor loads require consideration of both starting and running current.

Redundancy and emergency considerations also play important roles in feeder design. Critical facilities may employ dual feeders, automatic transfer switches, or separately derived systems to maintain power availability.

8.5 Common Feeder Types

Single-phase Feeders

Single-phase feeders are common in residential applications and smaller commercial buildings.

These typically consist of:

- Two hot conductors (for 240V single-phase).
- One neutral conductor.
- One equipment grounding conductor.

A typical residential application might include a 100A feeder from the main service panel to a detached garage subpanel, requiring properly sized conductors (typically 3 AWG copper for 100A), appropriate overcurrent protection, and a disconnecting means.

Three-phase Feeders

Three-phase feeders are common in commercial and industrial applications due to the higher power demands.

These configurations include:

- Three-phase, four-wire systems (three-phase conductors plus neutral)
- Three-phase, three-wire systems (three-phase conductors, no neutral)

They deliver more power with smaller conductors than equivalent single-phase systems. For example, a 200A, 480V three-phase feeder delivers approximately 166 kVA, while a 200A, 240V single-phase feeder delivers only 48 kVA—less than one-third the capacity.

The proper installation of feeders often presents unique challenges beyond basic sizing. Considerations like parallel conductors for high-current applications, appropriate derating for mutual heating, and the mechanics of pulling large conductors through complex raceway systems all require specialized knowledge and experience.

The journeyman exam typically includes questions on feeder calculations, including applying demand factors, sizing conductors, and determining appropriate overcurrent protection. Master these calculations, and you'll be well-prepared both for the exam and for designing effective electrical distribution systems.

8.6 Panel Boards and Switchgear

"That can't possibly be the main switchgear room," I thought as I stared at the tiny closet on the blueprint. The engineer had allocated barely enough space for the equipment, let alone the working clearances required by code. It wasn't the first time I'd seen this mistake, and I knew it wouldn't be the last. In electrical construction, panel boards and switchgear often are the flash points where design meets reality—and sometimes they clash spectacularly.

Panel boards and switchgear represent the control centers of electrical systems. These distribution hubs collect power from service entrances or feeders and distribute it through branch circuits to the furthest reaches of a building. While they might look like simple metal boxes filled with circuit breakers to the untrained eye, experienced electricians recognize them as sophisticated assemblies that require careful selection, installation, and maintenance.

The choices made at these distribution points cascade throughout the entire electrical system. Improper sizing, insufficient short-circuit ratings, or inadequate working space can create problems that persist for the lifetime of a building. Conversely, well-designed and properly installed equipment provides decades of reliable service with minimal maintenance.

Panel Boards

A panel board (or distribution panel) is the central hub for distributing electrical power throughout a building. It houses circuit breakers or fuses that protect individual circuits.

Panel boards serve as the nerve centers of electrical systems, receiving power from service equipment or feeders and distributing it to branch circuits. These critical components must be properly selected, installed, and maintained to ensure safe, reliable electrical systems. While often referred to generically as "panels" or "breaker boxes," panel boards vary significantly in design, features, and application.

Components of Panel Boards

Main Breaker: The primary overcurrent protection device for the entire panel. This device allows for disconnection of all power flowing through the panel and provides overcurrent protection for the panel's bus structure. The main breaker must be properly rated for both:

- The maximum load the panel will serve (ampacity rating)
- The maximum fault current available at its location (interrupting rating)

Not all panels include main breakers—some are designed as "main lug only" (MLO) panels, which receive protection upstream. These configurations are commonly used in multi-panel installations where a single main disconnect protects several downstream panels.

Branch Circuit Breakers: The source of protection for individual circuits extending to lights, receptacles, and other loads. These devices must be properly sized for the circuit conductors and the loads they serve. Modern panels accommodate various breaker types:

- Standard thermal-magnetic circuit breakers
- GFCI and AFCI protective devices
- Specialized breakers for equipment protection
- Multi-pole breakers for 240V or three-phase loads

The physical arrangement of these breakers connects to the internal bus structure, with the mounting configuration (plug-on or bolt-on) affecting both reliability and cost.

Bus Bars: Conductive strips that distribute power from the main breaker to the branch circuit breakers. These copper or aluminum bars provide structural support for circuit breakers while carrying substantial current. The bus bar design determines:

- The panel's maximum current rating
- The physical arrangement of breakers
- The configuration of phases (for three-phase panels)
- The short-circuit withstand rating

Additional components include neutral bars for grounded circuit conductors, equipment ground bars for grounding conductors, and the enclosure itself, which provides physical protection and mounting points.

Sizing and Selection

Determining the appropriate size of the panel board requires analysis of total load demand, number of circuits, and future expansion needs.

The ampacity rating must accommodate the calculated load plus a margin for future expansion. NEC Article 220 provides the methodology for load calculations based on:

- Square footage (for general lighting)
- Fixed appliance loads
- Motor loads
- Heating and cooling equipment
- Other specific loads

Beyond simple ampacity, panel selection involves evaluating:

Available Fault Current at the installation location must not exceed the panel's short-circuit current rating (SCCR). Modern panels typically range from 10,000 AIC to 200,000 AIC, with higher ratings required in locations closer to the utility service or large transformers.

Physical Space Requirements include both the number of circuit positions needed and the wall space required for mounting. A common mistake in residential applications is selecting a panel with insufficient circuit positions, requiring expensive add-ons later.

Environmental Conditions dictate the enclosure type, with options ranging from standard indoor enclosures (NEMA 1) to weatherproof (NEMA 3R) or corrosion-resistant versions for harsh environments.

Service Type (single-phase or three-phase) and voltage requirements directly affect panel selection. Three-phase panels have specific bus arrangements that must match the electrical service configuration.

Future Expansion should always factor into panel selection. The additional cost of a larger panel is minimal compared to the expense of adding panels later when space runs out.

The NEC addresses panel board requirements primarily in Article 408, and covers installation requirements, clearances, and protection. For residential applications, Article 210.11 specifies minimum circuit requirements that directly impact panel sizing. Commercial and industrial applications often involve more complex calculations based on diverse load types.

Panel boards represent a significant investment in both materials and labor. Selecting appropriate equipment that meets current needs while accommodating future growth prevents costly modifications and service interruptions down the road. As you develop expertise in this area, you'll learn to balance initial costs against long-term considerations, creating distribution systems that serve buildings effectively for decades.

8.7 Switchgear

Switchgear is used in high-voltage systems to control, protect, and isolate electrical equipment. It consists of switches, fuses, and circuit breakers.

While panel boards handle distribution for branch circuits, switchgear operates at a higher level in the electrical hierarchy. These robust assemblies manage larger power flows and provide more sophisticated protection and control capabilities. Switchgear represents the heavy artillery of electrical distribution—designed to handle massive power levels and interrupt dangerous fault currents safely.

Purpose of Switchgear

Switchgear protects electrical equipment from damage due to overloads, short circuits, or faults.

Beyond simple overcurrent protection, switchgear provides comprehensive power management:

- Isolation of electrical systems during maintenance or emergencies.
- Protection against ground faults and phase imbalances.
- Coordination with upstream and downstream protective devices.
- Monitoring and metering of electrical parameters.
- Control of power flow between different parts of an electrical system.

Modern switchgear often incorporates advanced protective relaying that can detect subtle abnormalities before they develop into catastrophic failures. These protection schemes might include differential relaying, zone interlocking, and communication with building management systems—capabilities far beyond basic circuit breakers.

8.8 Types of Switchgear

Low-voltage Switchgear

Low-voltage switchgear operates at 600V and below, encompassing the range used in most commercial and light industrial applications.

This equipment typically includes:

- Power circuit breakers with higher interrupting ratings than standard molded-case breakers.
- Draw-out construction that allows breakers to be removed while the switchgear remains energized.
- Compartmentalized design that separates bus, breaker, and cable sections for improved safety.
- Metering and monitoring capabilities for power management.

Low-voltage switchgear often serves as the main distribution point in commercial buildings, feeding downstream panel boards and motor control centers. Its robust construction and higher fault current ratings make it appropriate for service entrances and primary distribution.

Medium and High-voltage Switchgear

Medium-voltage switchgear (typically 1kV to 38kV) and high-voltage switchgear (above 38kV) serve industrial, utility, and power generation applications.

These sophisticated assemblies incorporate:

- Vacuum or SF6 gas-insulated breakers that can safely interrupt massive fault currents
- Elaborate safety interlocks to prevent access while energized
- Remote operation capabilities for hazardous environments
- Advanced protection relaying for discriminating between normal and fault conditions
- Arc-resistant construction to protect personnel if internal arcing occurs

Medium-voltage switchgear often serves as the interface between utility power and facility distribution systems in larger buildings and industrial complexes. The substantial size, cost, and complexity of this equipment require specialized knowledge for proper specification and installation.

8.9 Conduit Types and Raceway Fill Calculations

"You can't pull those conductors through that conduit," said the inspector, shaking his head at the overstuffed EMT run. "Look at the deformation on those wires." The frustrated apprentice beside me had spent hours trying to force too many conductors through an undersized raceway—a costly mistake that would require removing everything and starting over.

Conduit and raceway systems might not seem glamorous, but they are the backbone of most electrical installations. They protect conductors, provide physical support, and create safe routes through buildings. Selecting the right type for each application and correctly calculating fill requirements separates professionals from amateurs in our trade.

I've watched countless electricians learn the hard way that conduit sizing isn't something to eyeball. "That looks big enough" has derailed more projects than nearly any other assumption. Even experienced electricians can be surprised when the actual fill calculations reveal that their intuitive sizing was inadequate, especially when derating factors come into play.

The code provides specific guidelines for raceway applications and fill calculations for good reason. Forcing too many conductors into a raceway isn't just a code violation—it risks damaging insulation, creates excessive pulling tension, and generates heat that can degrade the entire installation over time. Mastering these calculations and understanding the proper applications for different raceway types will prevent costly errors and ensure your installations stand the test of time.

Types of Conduit

Conduit is a protective tubing used to house electrical wires. It protects from mechanical damage and helps organize the wiring system.

The selection of appropriate conduit types directly impacts installation durability, safety, and code compliance. Each type has specific advantages, limitations, and approved applications. While cost often influences selection, choosing the wrong conduit type for an application can lead to premature failure, code violations, and safety hazards.

Rigid Metal Conduit (RMC)

Rigid Metal Conduit represents the most robust and protective raceway option available. Made from galvanized steel, aluminum, or stainless steel, RMC provides superior protection against physical damage.

Key characteristics include:

- Thick walls (approximately 0.13 inches) that resist crushing and impact
- Threaded connections that create watertight and dustproof systems when properly installed
- Built-in grounding path through the metal construction
- Wide temperature tolerance and excellent fire resistance

RMC finds application in:

- hazardous locations (Class I, II, and III areas)
- areas subject to severe physical damage
- outdoor and wet locations
- service entrances and risers
- industrial environments with chemical exposure

While RMC offers maximal protection, its weight (approximately 0.85 pounds per foot for 3/4" steel RMC) makes installation labor-intensive and its cost significantly higher than lighter alternatives.

Intermediate Metal Conduit (IMC)

Intermediate Metal Conduit offers a middle ground between RMC's robustness and EMT's lighter weight. With wall thickness about 20% less than RMC, IMC provides substantial protection while reducing weight and material costs.

Notable features include:

- thinner walls than RMC (approximately 0.10 inches) but thicker than EMT
- threaded connections similar to RMC
- galvanized exterior and organic corrosion-resistant interior coating
- same thread dimensions as RMC, allowing interchangeable fittings

IMC is commonly used in:

- commercial and industrial applications
- areas with moderate physical damage concerns
- outdoor installations
- service entrances as a cost-effective alternative to RMC

The NEC permits IMC in almost all applications where RMC is allowed, which is what makes it an economical choice when maximum protection isn't required.

Electrical Metallic Tubing (EMT)

This is usually called "thin-wall" conduit. EMT is the most commonly used metallic raceway in commercial and institutional buildings. Its lighter weight and lower cost make it the go-to choice for most indoor installations.

EMT features:

- thin walls (approximately 0.05 inches for 3/4" EMT)
- compression or set-screw fittings rather than threaded connections
- galvanized steel or aluminum construction
- easy bending with hand tools or mechanical benders
- significant cost savings compared to RMC and IMC

EMT is ideal for:

- commercial office buildings
- schools and institutional facilities
- indoor exposed installations
- areas where moderate physical protection is adequate
- cost-sensitive projects with large conduit quantities

While versatile, EMT is not permitted in hazardous locations or areas subject to severe physical damage.

Flexible Metal Conduit (FMC)

Flexible Metal Conduit (often called "flex" or "Greenfield") consists of a spiral-wound metal strip that creates a flexible, protective pathway for conductors. This flexibility allows for:

- connections to vibrating equipment like motors
- installation around obstacles and irregular paths
- final connections in limited-access locations
- accommodation of building settlement or movement

FMC comes in standard flexible metal conduit and Liquidtight Flexible Metal Conduit (LFMC) varieties. LFMC adds a PVC jacket that provides watertight protection for wet locations or washdown areas.

Applications include:

- motor and equipment connections
- HVAC vibration isolation
- retrofit installations where rigid conduit would be impractical
- areas requiring movement or adjustment

The NEC limits FMC to 6 feet in most applications to prevent excessive voltage drop and potential damage from flexing.

Plastic Conduit (PVC)

Polyvinyl Chloride conduit offers corrosion resistance and economic advantages that make it popular for specific applications. PVC comes in various schedules (wall thicknesses), with Schedule 40 and Schedule 80 being most common in electrical applications.

PVC conduit provides:

- complete corrosion resistance
- lower cost than metallic alternatives
- lightweight installation (about 1/3 the weight of steel RMC)
- simple cutting and assembly with solvent cement
- excellent insulating properties

PVC is commonly used in:

- underground installations
- encased in concrete (especially Schedule 40)
- corrosive environments where metals would degrade
- direct burial service entrances
- agricultural applications

The non-metallic nature of PVC creates two significant limitations: it doesn't provide a grounding path, requiring a separate equipment grounding conductor, and it has poor performance in fires, generating toxic smoke and failing quickly in high temperatures.

Selecting the appropriate conduit type requires balancing protection needs, environmental conditions, code requirements, and budget constraints. The NEC addresses conduit application requirements in the dedicated article for each type (Articles 342-362), with specific limitations and uses permitted.

8.10 Raceway Fill Calculations

Raceway fill calculations determine how many conductors can safely fit inside a conduit without exceeding the recommended fill limit, which is critical for preventing overheating and ensuring code compliance.

Proper raceway ensures that you have adequate space for conductor cooling, which prevents insulation damage during installation, and allows for future maintenance or additions. These calculations represent one of the most frequently used and practical skills in day-to-day electrical work.

Purpose of Fill Calculations

The primary goal of raceway fill limitations is preventing conductor damage and overheating. When conductors are packed too tightly:

- Heat dissipation is reduced, potentially causing insulation damage
- Pulling tension increases dramatically, risking insulation damage during installation
- Conductor removal or replacement becomes difficult or impossible
- Conductor insulation can abrade against the raceway or other conductors

The NEC limits conduit fill to prevent these issues, with different percentages based on the number of conductors:

- One conductor: 53% maximum fill
- Two conductors: 31% maximum fill
- Three or more conductors: 40% maximum fill

These percentages represent the ratio of the combined cross-sectional area of all conductors to the internal cross-sectional area of the conduit.

NEC Guidelines

According to the National Electrical Code (NEC), conduit fill is governed by specific rules that help prevent overloading.

NEC Chapter 9, Tables 1 through 4 provide the necessary information for making these calculations:

- Table 1: Percent fill (as noted above)
- Table 4: Dimensions and area of conduits and tubing
- Table 5: Dimensions of insulated conductors

For standard installations with four or fewer conductors, the NEC also provides Tables C.1 through C.12 (Annex C) that give the maximum number of conductors permitted in various conduit types—eliminating the need for calculations in many common scenarios.

Article 310.15(B)(3)(a) adds another crucial consideration: conductor ampacity derating when more than three current-carrying conductors run in a single raceway. This can impact both conductor sizing and raceway selection.

Calculation Methodology

To determine proper conduit size, follow these steps:

1. Identify all conductors that will run in the raceway, including their size and insulation type.
2. Determine the cross-sectional area of each conductor from Table 5 in Chapter 9.
3. Calculate the total cross-sectional area by adding all the individual conductor areas.
4. Divide this total by the appropriate fill percentage (0.40 for three or more conductors).
5. Select a conduit size from Table 4 with an internal area greater than or equal to this calculated value.

For example, calculating the minimum conduit size for six #10 THHN conductors:

- Area of one #10 THHN = 0.0211 square inches (from Table 5).
- Total area = 6 × 0.0211 = 0.1266 square inches.
- Required conduit area = 0.1266 ÷ 0.40 = 0.3165 square inches.
- From Table 4, 3/4" EMT has an internal area of 0.353 square inches.
- Therefore, 3/4" EMT is the minimum size required
- Special installations like motor circuits may have specific provisions.

8.11 Quiz: Testing Your Knowledge of Wiring and Circuitry Applications

You've absorbed a lot of information about branch circuits, feeders, panel boards, switchgear, and conduit systems. Let's see how well you can apply these concepts to real-world scenarios. This quiz simulates the types of questions you might encounter on your journeyman exam while reinforcing the practical knowledge you'll need in the field.

Some questions will test your recall of specific requirements, while others will challenge you to work through calculations similar to those you'll perform on actual jobs. Take your time, read carefully, and remember that the correct answer often depends on specific details in the question.

Multiple-Choice Questions

1. **Which of the following branch circuits is required by the NEC to be dedicated (not serving any other outlets or loads)?**

    a) Circuit for refrigerator in a kitchen
    b) Circuit for bathroom receptacles
    c) Circuit for laundry receptacles
    d) Circuit for garage door opener

2. **What is the minimum size EMT conduit required for four #10 THHN conductors?**

    a) 1/2 inch
    b) 3/4 inch
    c) 1 inch
    d) 1-1/4 inch

3. **In a 120/240V single-phase panel, what is the purpose of the neutral bus bar?**
   a) Provides a connection point for equipment grounding conductors
   b) Provides a connection point for grounded (neutral) conductors
   c) Provides overcurrent protection for branch circuits
   d) Provides a path for fault current to return to the source

4. **Which of the following is NOT a proper application for Rigid Metal Conduit (RMC)?**
   a) Hazardous locations classified as Class I, Division 1
   b) Underground installations directly buried in earth
   c) Embedded in concrete below ground level
   d) Data processing centers with sensitive electronic equipment

5. **According to the NEC, what is the maximum number of general-purpose circuit breakers permitted in a lighting and appliance branch-circuit panelboard?**
   a) 30
   b) 40
   c) 42
   d) 60

6. **A three-phase, four-wire feeder supplies a 200A panel. If the feeder conductors are copper with THHN insulation, what is the minimum size required?**
   a) 2/0 AWG
   b) 3/0 AWG
   c) 4/0 AWG
   d) 250 kcmil

7. **True or False: NEC permits Type NM (Romex) cable to be used in a commercial office building that has a finished height of two stories.**
   a) True
   b) False

8. **Which of the following is a characteristic of a main lug only (MLO) panel?**
   a) It has no main breaker
   b) It has no neutral bus
   c) It cannot be used as a subpanel
   d) It must always be installed as service equipment

9. **The NEC requires a minimum of how many small-appliance branch circuits for a dwelling unit kitchen?**

   a) 1
   b) 2
   c) 3
   d) 4

10. **Which type of conduit is most appropriate for an installation in a corrosive chemical environment?**

    a) EMT with compression fittings
    b) Rigid Metal Conduit
    c) PVC Schedule 80
    d) Flexible Metal Conduit

11. **True or False: Three-way switches require a neutral at each switch location according to the current NEC.**

    a) True
    b) False

12. **What is the primary difference between switchgear and switchboards?**

    a) Switchgear uses fuses while switchboards use circuit breakers
    b) Switchgear has higher interrupting ratings and compartmentalized construction
    c) Switchboards are used only in residential applications
    d) Switchgear is limited to low-voltage applications

13. **What is the maximum number of bends allowed between pull points in a conduit run?**

    a) 180 degrees
    b) 270 degrees
    c) 360 degrees
    d) No limit as long as conductors can be pulled without damage

14. **In a 277/480V system, what color is typically used to identify the Phase C conductor?**

    a) Black
    b) Red
    c) Blue
    d) Yellow

15. **True or False: A panel board used as service equipment must have a main disconnecting means.**

    a) True
    b) False

Problem-Solving Questions

16. **A 100-foot run of conduit will contain the following THHN conductors: six #12, four #10, and two #8. Calculate the minimum size EMT conduit required.**

17. **A single-family dwelling has the following loads:**

    * General lighting load: 3,000 VA
    * Small appliance circuits: Two 20A circuits
    * Laundry circuit: One 20A circuit
    * Dishwasher: 1,200 VA
    * Garbage disposal: 900 VA
    * Electric range: 12 kW
    * Electric water heater: 4,500 VA
    * Central AC: 24A at 240V

    Calculate the minimum size service required for this dwelling.

18. **A feeder supplies a panel with the following loads:**

    * 18 kW of continuous lighting load at 277V
    * 22 kW of receptacle load (non-continuous) at 277V
    * 30 kW of air conditioning load at 480V, 3-phase

    Calculate the minimum ampacity required for the feeder conductors.

19. **A control panel contains the following equipment in a common enclosure:**

    * Main disconnect: 100A
    * Six 20A, 1-pole circuit breakers
    * Three 30A, 2-pole circuit breakers
    * Two 50A, 3-pole circuit breakers

    How many poles total are in this panel, and does it comply with the NEC limit of 42 poles for a lighting and appliance branch-circuit panelboard?

20. **A run of 1-inch EMT contains four current-carrying conductors: two #6 THHN and two #8 THHN. What percentage of conduit fill do these conductors represent, and is this installation code compliant?**

Answer Key and Explanations

1. **Answer: c) Circuit for laundry receptacles**

   NEC 210.11(C)(2) requires at least one 20-ampere branch circuit for laundry receptacles. This circuit shall have no other outlets, making it a dedicated circuit.

2. **Answer: a) 1/2 inch**

   Four #10 THHN conductors have a total area of: 4 × 0.0211 = 0.0844 square inches. With 40% fill for four conductors, the required conduit area is 0.0844 ÷ 0.4 = 0.211 square inches. 1/2-inch EMT has an area of 0.213 square inches, making it adequate.

3. **Answer: b) Provides a connection point for grounded (neutral) conductors**

   The neutral bus bar in a panel is specifically for terminating the grounded (neutral) conductors. Equipment grounding conductors connect to a separate ground bus.

4. **Answer: d) Data processing centers with sensitive electronic equipment**

   While RMC is suitable for hazardous locations and underground installations, it's not specifically required for data centers. Many data centers avoid ferrous metal conduits to minimize electromagnetic interference.

5. **Answer: c) 42**

   NEC 408.54 limits lighting and appliance branch-circuit panelboards to 42 overcurrent devices.

6. **Answer: b) 3/0 AWG**

   For a 200A feeder, copper THHN conductors must have an ampacity of at least 200A. According to Table 310.16, 3/0 AWG copper has an ampacity of 200A at 75°C.

7. **Answer: b) False**

   NEC 334.10 restricts Type NM cable in commercial buildings to structures not exceeding three floors above grade. However, many local amendments further restrict its use in commercial applications regardless of height.

8. **Answer: a) It has no main breaker**

   A main lug only (MLO) panel does not have a main breaker and receives its overcurrent protection from an upstream device.

9. **Answer: b) 2**

   NEC 210.11(C)(1) requires a minimum of two 20-ampere small-appliance branch circuits for receptacle outlets in the kitchen, pantry, breakfast room, and dining room.

10. **Answer: c) PVC Schedule 80**

    PVC Schedule 80 provides excellent resistance to most chemicals and is the appropriate choice for corrosive environments. Metallic conduits would deteriorate in such conditions.

11. **Answer: a) True**

    The 2020 NEC 404.2(C) requires a neutral conductor at switch locations where electronic lighting control devices or switches with pilot lights are installed, even if not needed for the basic 3-way switching function.

12. **Answer: b) Switchgear has higher interrupting ratings and compartmentalized construction**

    Switchgear features higher interrupting ratings, compartmentalized construction for enhanced safety, and often uses draw-out circuit breakers. Both use circuit breakers, and both can be used in various applications at different voltage levels.

13. **Answer: c) 360 degrees**

    NEC 358.26 (for EMT) and similar sections for other conduit types limit the total bends between pull points to 360 degrees (four 90-degree bends).

14. **Answer: d) Yellow**

    While practices vary by region, the common color code for 480V systems is brown-orange-yellow for phases A, B, and C, respectively.

15. **Answer: a) True**

    NEC 230.70 requires a service disconnecting means for all service equipment. A panel serving as service equipment must have a main disconnect.

16. **Answer: 1-inch EMT**

Conductor areas:

* Six #12 THHN: 6 × 0.0133 = 0.0798 square inches
* Four #10 THHN: 4 × 0.0211 = 0.0844 square inches
* Two #8 THHN: 2 × 0.0366 = 0.0732 square inches
* Total area: 0.2374 square inches

At 40% fill, required conduit area = 0.2374 ÷ 0.4 = 0.5935 square inches. From Table 4, 1-inch EMT has an area of 0.864 square inches, making it the minimum size required.

17. **Answer: 100 amperes**

Load calculation:

* General lighting: 3,000 VA
* Small appliance circuits: 2 × 1,500 VA = 3,000 VA
* Laundry circuit: 1,500 VA
* First 10,000 VA at 100%: 10,000 VA
* Remaining load (3,000 + 1,200 + 900 + 12,000 + 4,500) = 21,600 VA × 40% = 8,640 VA
* AC: 24A × 240V = 5,760 VA × 100% = 5,760 VA
* Total: 10,000 + 8,640 + 5,760 = 24,400 VA

At 240V: 24,400 ÷ 240 = 101.7A Standard size: 110A, but 100A is commonly used as service size.

18. **Answer: 98 amperes**

* Continuous lighting: 18 kW × 125% = 22.5 kW
* Receptacle load: 22 kW (non-continuous)
* AC load: 30 kW
* Total load: 22.5 + 22 + 30 = 74.5 kW

For 480V, 3-phase: I = P ÷ (√3 × V × PF) Assuming PF = 0.9: I = 74,500 ÷ (1.732 × 480 × 0.9) = 98 amperes

19. **Answer: 26 poles, yes, compliant**

* Main disconnect: 2 poles (assuming single-phase)
* Six 1-pole breakers: 6 poles
* Three 2-pole breakers: 6 poles
* Two 3-pole breakers: 6 poles
* Total: 26 poles

This is below the 42-pole limit for lighting and appliance panelboards.

20. **Answer: 30.8%, yes compliant** Conductor areas:

* Two #6 THHN: 2 × 0.0507 = 0.1014 square inches
* Two #8 THHN: 2 × 0.0366 = 0.0732 square inches
* Total area: 0.1746 square inches

1-inch EMT area: 0.566 square inches. Fill percentage: (0.1746 ÷ 0.566) × 100 = 30.8.% For four conductors, maximum fill is 40%, so this installation is code compliant.

9

Grounding and Bonding Principles

Proper grounding can mean the difference between life and death. That statement might sound dramatic, but every experienced electrician has encountered installations where compromised grounding created hazards invisible to untrained eyes. A loose connection, an incorrectly sized conductor, or a misapplied bonding jumper can transform an otherwise proper electrical system into a potential tragedy waiting to happen.

Grounding and bonding represent the foundation of electrical safety, yet they remain among the most misunderstood aspects of our trade. While most electricians can recite that "grounding provides a path for fault current" or "bonding connects metal parts," fewer truly grasp how these systems work together to create safe electrical installations.

In this chapter, we'll explore the principles that make grounding and bonding work, examine how these systems protect both people and equipment, and address the common misconceptions that lead to dangerous installations. Whether you're working on residential services, commercial distribution systems, or industrial power supplies, understanding these fundamentals will help you create safer electrical systems and identify hazardous conditions that others might miss.

The difference between an ordinary electrician and an exceptional one often lies in their understanding of these principles. Code-minimum installations might pass inspection, but truly safe systems require the deeper knowledge we'll develop here.

9.1 NEC Grounding Requirements

Grounding represents one of the most critical safety elements in any electrical system, yet it's frequently misunderstood and improperly implemented. Understanding both why we ground electrical systems and how to do it correctly according to code can save lives and prevent property damage.

Purpose of Grounding

Safety: Grounding is essential to provide a safe path for fault currents to flow, preventing the potential for electrical shock or fire.

When a fault occurs in an electrical system, current needs a definite, low-impedance path back to the source to facilitate enough current flow to operate overcurrent devices quickly. Without proper grounding, fault current might take unpredictable paths—potentially through people or combustible materials.

Effective grounding creates a deliberate, controlled path for this current, ensuring that circuit breakers or fuses operate promptly to disconnect faulted circuits. This fundamental safety function protects both people and property from the dangers of uncontrolled electrical energy.

System Protection: Grounding also helps to stabilize the voltage levels within an electrical system by providing a common reference point for all electrical components.

Beyond safety, grounding serves several system functions include:

- limits overall system voltage with respect to earth during normal operation
- provides a zero-voltage reference point for control and instrumentation systems
- helps minimize electromagnetic interference in sensitive electronic equipment
- facilitates lightning protection by providing paths for lightning energy to dissipate

These system benefits enhance reliability and performance while complementing the primary safety functions of the grounding system.

9.2 NEC Grounding Methods

"Just pound in the ground rod and hook up the wire," the old-timer told me on my first service change. When I asked about the water pipe, building steel, or concrete-encased electrode, he waved me off. "Nobody checks all that—one rod has worked fine for forty years."

That dangerous misconception represents exactly why the NEC devotes an entire article to proper grounding methods. A single ground rod might appear to work, but during an actual fault condition, that inadequate connection to earth could mean the difference between a safely cleared fault and a deadly energized case.

There are very specific methods for creating effective grounding systems that go far beyond simply "connecting to ground." These requirements create redundant, reliable connections to earth that remain effective even during fault conditions or after years of environmental exposure.

Grounding Electrode System

A grounding electrode system is a collection of conductors that connect the electrical system to the earth. Key components include ground rods, ground plates, and metal water pipes.

The NEC requires the use of all available electrodes at a building or structure to form a grounding electrode system. These electrodes must be bonded together, creating redundancy that ensures the ground connection remains effective even if one electrode becomes compromised.

Common grounding electrodes include:

- **Metal Underground Water Pipe**: A metal water pipe in direct contact with the earth for at least 10 feet qualifies as a grounding electrode. However, NEC 250.53(D)(2) prohibits using this as the sole electrode—it must be supplemented with another type. This requirement recognizes that water pipes may be replaced with non-metallic materials, potentially eliminating the ground connection without warning.
- **Building Steel**: Metal in direct contact with the earth or encased in concrete that touches the earth can serve as a grounding electrode. Structural steel often provides an excellent low-impedance path to earth due to its large contact area.
- **Concrete-Encased Electrode (Ufer Ground)**: It was named after Herbert Ufer, who developed it during World War II, and consists of at least 20 feet of bare copper conductor (not smaller than #4 AWG) encased in concrete that contacts the earth. The concrete's alkaline environment and large surface area make this one of the most effective electrodes available.
- **Ground Rods**: Typically 8 feet long and 5/8 inch in diameter, they must be driven until at least 8 feet contacts the soil. When a single rod doesn't achieve 25 ohms resistance to earth, a second rod must be installed at least 6 feet away. The NEC doesn't require testing if two rods are installed.
- **Ground Rings**: A bare copper conductor (at least #2 AWG) encircling the building at a depth of at least 30 inches qualifies as an electrode. These are particularly effective for buildings with extensive perimeters.

Grounding Conductors

These conductors connect the system's neutral to the grounding electrode system. The size and type of the conductor depend on the system's amperage and the type of installation.

The NEC recognizes several distinct types of grounding conductors, each with specific functions:

Grounding Electrode Conductor (GEC): Connects the grounded conductor (usually the neutral) or the equipment grounding conductor to the grounding electrode system. NEC 250.66 specifies sizing requirements based on the size of the largest service-entrance or derived system conductors. For standard services with copper service conductors, common GEC sizes include:

- Up to 2 AWG service conductors: 8 AWG copper GEC
- 1/0 to 3/0 service conductors: 6 AWG copper GEC
- Over 3/0 through 350 kcmil: 4 AWG copper GEC
- Over 350 kcmil: 2 AWG copper GEC

Importantly, when connecting to ground rods or pipes, the GEC need not be larger than 6 AWG copper, regardless of service size. For concrete-encased electrodes, 4 AWG copper is the maximum required size.

Equipment Grounding Conductor (EGC): Connects non-current-carrying metal parts of equipment to the system grounded conductor, the grounding electrode conductor, or both at the service equipment. NEC Table 250.122 specifies sizes based on the rating of the overcurrent device protecting the circuit, not the actual load current. For standard branch circuits:

- 15A circuit: 14 AWG copper EGC
- 20A circuit: 12 AWG copper EGC
- 60A circuit: 10 AWG copper EGC
- 100A circuit: 8 AWG copper EGC

Main Bonding Jumper: Connects the grounded system conductor (neutral) to the equipment grounding conductor(s) and the grounding electrode conductor at service equipment. This critical connection ties the entire grounding system together at a single point, preventing parallel current paths on grounding conductors.

System Bonding Jumper: Serves the same function as the main bonding jumper but at separately derived systems like transformers or generators.

Bonding

Bonding is when you connect the metal parts of an electrical system to the ground to ensure they share a common electrical potential, reducing the risk of shock or fire.

Effective bonding ensures that all conductive objects that could become energized remain at the same potential, preventing dangerous voltage differences. The NEC specifies several bonding requirements:

Service Equipment Bonding (250.92): Requires bonding of service raceways, cable trays, cablebus framework, and service cable armor or sheaths. These connections must be made with listed bonding devices, exothermic welding, or other approved means.

Equipment Bonding (250.96): Ensures that electrical continuity exists between all metal raceways, cables, enclosures, and equipment. This continuity creates a low-impedance path for fault current.

Bonding of Piping Systems and Exposed Structural Steel (250.104): Requires bonding of metal water piping, metal gas piping, and exposed structural steel that is likely to become energized. These bonding connections prevent dangerous potential differences during fault conditions or lightning strikes.

Bonding Methods (250.8, 250.70): Specifies acceptable methods for making bonding connections, including:

- listed pressure connectors
- terminal bars
- exothermic welding
- listed clamps
- other listed means

Proper bonding doesn't just prevent shock hazards—it ensures that fault currents are high enough to operate overcurrent devices quickly, minimizing the duration of fault conditions and associated fire risks.

The key to effective grounding and bonding lies in understanding that it's a complete system, not just individual components. Each element—electrodes, conductors, connections, and bonded equipment—plays a crucial role in maintaining safety throughout the electrical installation.

9.3 Grounding Conductor Sizing

The size of grounding conductors isn't arbitrary—it's carefully calculated to handle fault currents safely without failing. An undersized grounding conductor might work fine under normal conditions but could fail catastrophically during a fault, rendering the entire grounding system ineffective precisely when it's most needed.

Determining Size

The size of the grounding conductor is critical for safely dissipating fault currents. It is determined by the size of the largest ungrounded conductor and the type of system.

The NEC distinguishes between different types of grounding conductors, each with its own sizing requirements:

Grounding Electrode Conductor (GEC) sizing is covered in NEC 250.66 and depends on the size of the largest service-entrance conductor or equivalent area for parallel conductors. For copper conductors, common sizing requirements include:

Service Conductor Size	Copper GEC Size	Aluminum GEC Size
2 AWG or smaller	8 AWG	6 AWG
1/0 to 3/0 AWG	6 AWG	4 AWG
Over 3/0 to 350 kcmil	4 AWG	2 AWG
Over 350 to 600 kcmil	2 AWG	1/0 AWG
Over 600 to 1100 kcmil	1/0 AWG	3/0 AWG
Over 1100 kcmil	2/0 AWG	4/0 AWG

Table 2: Common Sizing Requirements for Copper Conductors

Important exceptions modify these requirements for specific electrode types:

- When connecting to a ground rod or pipe electrode, the GEC need not be larger than 6 AWG copper or 4 AWG aluminum
- For a concrete-encased electrode, the GEC need not be larger than 4 AWG copper
- For a ground ring, the GEC need not be larger than the conductor used in the ground ring

Equipment Grounding Conductor (EGC) sizing follows NEC Table 250.122 and is based on the rating of the overcurrent device protecting the circuit, not the circuit conductors themselves. Common EGC sizes include:

Circuit Breaker/Fuse Size	Copper EGC Size	Aluminum EGC Size
15A	14 AWG	12 AWG
20A	12 AWG	10 AWG
30-40A	10 AWG	8 AWG
60A	10 AWG	8 AWG

Circuit Breaker/Fuse Size	Copper EGC Size	Aluminum EGC Size
100A	8 AWG	6 AWG
200A	6 AWG	4 AWG
300A	4 AWG	2 AWG
400A	3 AWG	1 AWG
500A	2 AWG	1/0 AWG
600A	1 AWG	2/0 AWG
800A	1/0 AWG	3/0 AWG
1000A	2/0 AWG	4/0 AWG
1200A	3/0 AWG	250 kcmil
1600A	4/0 AWG	350 kcmil
2000A	250 kcmil	400 kcmil
2500A	350 kcmil	600 kcmil
3000A	400 kcmil	600 kcmil
4000A	500 kcmil	800 kcmil
5000A	700 kcmil	1200 kcmil
6000A	800 kcmil	1200 kcmil

Table 3: Equipment Grounding Conductor Sizing Guide

When ungrounded conductors are increased in size for voltage drop or other reasons, the EGC must be proportionally increased as well. This requirement, found in 250.122(B), ensures the equipment grounding path maintains appropriate current-carrying capacity relative to the phase conductors.

Main Bonding Jumper sizing follows NEC 250.28 and must have at least 12.5% of the area of the largest phase conductor. In practice, main bonding jumpers are typically sized the same as the GEC, creating a continuous path of equivalent capacity.

9.4 Common Grounding Mistakes and How to Avoid Them

I am familiar with scenarios where a $30,000 piece of medical equipment was destroyed, where a data center with mysterious computer crashes, and a homeowner who received shocks from his kitchen faucet. What did these situations have in common? Grounding mistakes. Despite appearing simple in concept, proper grounding and bonding remain among the most frequently misunderstood and incorrectly implemented aspects of electrical work.

These mistakes rarely show themselves during normal operation. Systems with serious grounding deficiencies often function normally for years—until a fault occurs, equipment malfunctions, or someone gets hurt. By then, the damage is done. What makes these errors particularly dangerous is that they often remain invisible to casual inspection, hiding behind walls and under covers until a critical situation exposes the deficiency.

In my decades in the field, I've encountered the same fundamental grounding errors repeatedly, across projects ranging from residential services to industrial facilities. Recognizing these common mistakes—and knowing how to avoid them—can save you from callbacks, equipment damage, and potentially dangerous situations. Let's examine the most frequent grounding errors and how to prevent them in your installations.

Mistake 1: Incorrect Grounding Conductor Size

Problem: Sizing the grounding conductor too small can result in insufficient fault current dissipation, potentially causing damage to electrical equipment or creating a safety hazard.

I once investigated a burned-up service panel in a small commercial building. The main breaker had failed to trip during a major fault, resulting in melted conductors and a near-fire. The culprit? A #8 AWG grounding electrode conductor serving a 400-amp service. This drastically undersized conductor created too much impedance in the ground path, preventing proper fault clearing.

Undersized grounding conductors may carry normal leakage currents without issue, but when a ground fault occurs, they can:

- Fail to carry enough current to trip overcurrent devices quickly
- Overheat and potentially break, destroying the ground path entirely
- Create excessive voltage on equipment enclosures during faults
- Melt or burn, potentially causing fires or damaging equipment

This mistake occurs most frequently with grounding electrode conductors (GECs) and equipment grounding conductors (EGCs) for larger feeders and services. Many electricians mistakenly believe that "ground wires don't carry current," failing to understand that during a fault, these conductors must carry the entire fault current.

Solution: Refer to the NEC tables for proper grounding conductor size based on system amperage and wire material (copper or aluminum).

To avoid this mistake:

- Always consult NEC Table 250.66 for grounding electrode conductor sizing based on service or feeder size.
- Use Table 250.122 for equipment grounding conductor sizing based on overcurrent protection rating.
- Remember that increased conductor sizing for voltage drop requires proportional increase of equipment grounding conductors (250.122(B)).
- Don't assume a ground wire "doesn't need to be very big" because it normally doesn't carry current.
- Be particularly careful when sizing grounding conductors for parallel conductor installations.

For critical installations, consider exceeding the minimum code requirements. The cost difference between a code-minimum grounding conductor and the next size up is typically minimal compared to the safety margin it provides.

When inspecting existing installations, look for signs of undersized grounding conductors, particularly in older systems installed under previous code editions. Upgrading these conductors during renovations or service changes can eliminate potentially dangerous conditions before they cause problems.

Mistake 2: Failing to Bond Non-Current-Carrying Metal Parts

Problem: Failure to bond metallic parts such as conduit, electrical boxes, and grounding rods can result in a dangerous difference in potential, leading to shock hazards.

This mistake often occurs because electricians don't fully understand the purpose of bonding. While many do recognize that enclosures and raceways need to be connected to ground, they may miss other metal components that require bonding or fail to create proper bonding connections.

When metal parts aren't properly bonded, they can become "islands" at different electrical potentials. During a fault condition, this potential difference can be hundreds of volts—more than enough to cause serious injury or death if a person simultaneously contacts two metal parts at different potentials.

Common bonding failures include:

- Missing bonding jumpers around concentric knockouts in panels
- Failing to bond metal water piping systems
- Using reducing washers without bonding jumpers for different-sized conduits
- Neglecting to bond supplementary grounding electrodes together
- Failing to bond CSST gas piping as required by manufacturer instructions
- Not maintaining continuity across flexible conduit connections
- Omitting bonding bushings on service conduits entering metal enclosures

I once investigated a commercial kitchen where employees reported occasional shocks when touching both the sink and a nearby appliance. The culprit was a water pipe replacement that had introduced a plastic section, interrupting the bonding path for the metallic water piping system. During certain conditions, different sections of metal pipe developed different potentials, creating a shock hazard.

Solution: Always ensure that all metal parts are properly bonded to the grounding system, following NEC guidelines for bonding.

To avoid this mistake:

- Create a continuous bonding path for all metal enclosures, raceways, and equipment.
- Install bonding jumpers around concentric knockouts and where continuity might be compromised.
- Bond all metal piping systems that may become energized (water, gas, if CSST).
- Use appropriate bonding fittings for flexible connections.
- Ensure that all grounding electrodes are bonded together per NEC 250.50.
- Verify continuity with testing when unsure about connections.

Pay particular attention to NEC 250.92 for bonding at service equipment, 250.96 for general bonding of metal raceways and enclosures, and 250.104 for bonding of piping systems and exposed structural steel.

Remember that proper bonding requires a low-impedance path that can safely carry fault current. Connections must be made with listed devices, appropriate pressure connectors, or exothermic welding to ensure that they remain intact during fault conditions. When retrofitting or maintaining existing systems, be particularly vigilant about maintaining bonding continuity when replacing components. A water pipe replacement, panel upgrade, or equipment modification that interrupts bonding paths can introduce hazards into previously safe systems.

Mistake 3: Inadequate Grounding of Electrical Equipment

Problem: Electrical devices like appliances, power panels, and outlets need to be adequately grounded to prevent the risk of electrical shock or fire.

Inadequate equipment grounding is a common and dangerous electrical violation. When equipment cases aren't properly grounded, a simple insulation failure can energize metal surfaces at line voltage, creating serious shock hazards and potential fire risks.

This problem manifests in several ways:

- Missing or damaged equipment grounding conductors.
- Loose connections at grounding terminals.
- Using non-metallic boxes for metal devices without proper grounding provisions.
- Improper connections that rely on device yokes for grounding.
- Painted surfaces interfering with proper ground connections.
- Ungrounded metal enclosures and raceways.
- Equipment grounding conductors connected to neutral instead of ground bus in subpanels.

The consequences can be severe. I once investigated a case where a homeowner received a serious shock from a washing machine. The equipment grounding conductor had worked loose from the receptacle terminal, leaving the machine ungrounded. When an internal fault energized the machine's case, there was no low-impedance path to clear the fault—instead, the homeowner's body provided the path to ground.

This issue is particularly dangerous with appliances that connect to both electrical systems and water pipes, like dishwashers, washing machines, and water heaters. The water connection provides an excellent path to ground, potentially directing fault current through anyone who touches the appliance.

Solution: Use proper grounding techniques when installing equipment, ensuring that devices with metal enclosures are correctly grounded.

To avoid this mistake:

- Verify that equipment grounding conductors are properly sized per NEC 250.122.
- Torque all grounding connections to manufacturer specifications.
- Use grounding pigtails in metal boxes rather than relying on device yokes.
- Ensure proper bonding between metal boxes and conduit entries.
- Remove paint or other insulating materials from grounding connection points.
- Maintain separate ground and neutral buses in subpanels.
- Use appropriate fittings to maintain grounding continuity with flexible connections.
- Test ground continuity and resistance before energizing equipment.

Pay particular attention to Article 250, Part VI (Equipment Grounding and Equipment Grounding Conductors) for specific requirements. NEC 250.118 lists recognized equipment grounding conductor types, while 250.134 addresses the grounding of equipment fastened in place or connected by permanent wiring methods.

For cord-and-plug connected equipment, NEC 250.114 specifies which equipment requires grounding. Generally, all equipment with metal cases used in wet or damp locations, or within reach of grounded surfaces, requires grounding.

When working on older installations, be particularly vigilant about equipment grounding. Many older systems were installed before current grounding requirements were established, or have had grounding provisions compromised through years of modifications. Upgrading these systems to current standards should be recommended whenever possible.

Mistake 4: Incorrect Use of Grounding Electrodes

Problem: Not using the correct grounding electrodes (ground rods, plates, or water pipes) compromises the effectiveness of the grounding system.

This mistake takes several forms, from using inadequate electrodes to failing to utilize all available electrodes as required by code. The NEC is specific about grounding electrode requirements, yet this area sees frequent violations.

Common errors include:

- Installing only a single ground rod without verifying the 25-ohm resistance requirement.
- Failing to connect to available building steel, water pipes, or concrete-encased electrodes.
- Driving ground rods insufficiently deep or in unsuitable soil conditions.
- Making improper connections to grounding electrodes that corrode over time.
- Using grounding electrodes that don't meet NEC requirements for material or dimensions.
- Failing to bond multiple electrodes together as required by NEC 250.50.

Solution: Refer to the NEC's grounding electrode section and ensure you use the appropriate electrodes for the installation location and electrical system size.

To avoid this mistake:

- Always use all available grounding electrodes as required by NEC 250.50.
- When installing ground rods, ensure they have at least 8 feet of contact with soil.
- Either verify the 25-ohm resistance requirement for a single rod or install two rods at least

6 feet apart.
- Make connections to electrodes using listed clamps or exothermic welding.
- Ensure proper sizing of grounding electrode conductors per NEC 250.66.
- Pay particular attention to concrete-encased electrodes (Ufer grounds), which often provide superior grounding performance.

The NEC requires using all the following grounding electrodes that exist at each building or structure:

- Metal underground water pipes in direct contact with earth for at least 10 feet.
- Metal in-ground support structure (effectively grounded structural metal).
- Concrete-encased electrodes (at least 20 feet of bare copper conductor or steel reinforcing bars/rods not less than 1/2 inch in diameter, encased in at least 2 inches of concrete near the bottom of a foundation).
- Ground rings (at least 20 feet of bare copper conductor not smaller than 2 AWG buried at least 30 inches deep encircling the building).

If none of these electrodes are available, you must install one or more of the following:

- Ground rods (at least 8 feet long, 5/8 inch diameter for steel or 1/2 inch for copper or copper-clad)
- Plate electrodes (at least 2 square feet of surface exposed to soil)
- Other listed electrodes

When selecting and installing grounding electrodes, consider not just code compliance but actual effectiveness. Soil conditions, moisture levels, and seasonal variations can significantly affect grounding performance. In critical installations, periodic testing of ground resistance can help ensure continued effectiveness.

Mistake 5: Overlooking Grounding in Wet or Hazardous Locations

Problem: Wet or hazardous locations often require additional grounding considerations to prevent corrosion and maintain safe electrical operation.

Standard grounding methods that work well in dry, clean environments can rapidly deteriorate in wet, corrosive, or hazardous locations. This deterioration often occurs invisibly, compromising safety systems until a fault condition reveals the deficiency—often with dangerous consequences.

In wet locations, ordinary connections can corrode rapidly, particularly when dissimilar metals create galvanic action. A ground clamp that performs perfectly in a dry basement might fail within months when exposed to moisture, chemicals, or salt air.

Hazardous locations present additional challenges. In these environments, proper grounding isn't just about shock protection—it's also critical for preventing sparks that could ignite flammable gases, vapors, or combustible dust. A compromised grounding system in a hazardous location can lead to catastrophic explosions or fires.

I once investigated a marina where a swimmer reported receiving shocks while near a boat dock. The dock's electrical system had been grounded using standard materials and methods, but the constant exposure to water and salt had severely corroded the connections. What appeared visibly intact actually provided minimal grounding protection, creating a dangerous situation for anyone in the water nearby.

In another case, a grain processing facility experienced a dust explosion after static electricity ignited combustible grain dust. Investigation revealed that the facility's equipment grounding system had degraded over time, allowing static charges to accumulate without a proper path to ground.

Solution: Follow the specific NEC guidelines for grounding systems in wet or corrosive environments, including the use of corrosion-resistant materials.

To avoid this mistake:

- Use copper or stainless steel grounding components in wet or corrosive locations.
- Apply anti-oxidant compounds to all connections exposed to moisture.
- Select proper grounding connector types rated for the specific environment.
- Implement regular inspection programs for grounding systems in harsh environments.
- Use exothermic welding rather than mechanical connections where possible.
- Consider supplementary grounding methods for critical equipment in harsh conditions.
- Install additional equipotential bonding in areas like swimming pools, marinas, and agricultural buildings.

For hazardous locations, follow the specific requirements in NEC Articles 500-517, which address various classes and divisions of hazardous environments. These articles provide detailed requirements for grounding and bonding that go beyond standard provisions.

Pay particular attention to:

- NEC 406.9 for receptacles in wet locations.
- NEC 680 for swimming pools, spas, and similar installations.
- NEC 682 for natural and artificially made bodies of water.
- NEC 547 for agricultural buildings.
- NEC 501-503 for Class I, II, and III hazardous locations.

In wet locations, all connections should be made using connectors listed for wet locations, properly sealed against moisture intrusion, and installed to prevent water accumulation. Consider periodic testing of ground resistance in critical installations to detect deterioration before it leads to safety hazards.

Remember that grounding systems in harsh environments require ongoing maintenance and inspection. Even properly installed systems can degrade over time, particularly when exposed to moisture, chemicals, temperature extremes, or physical stress.

9.5 Quiz: Testing Your Knowledge of Grounding and Bonding Principles

Proper grounding and bonding are among the most crucial yet frequently misunderstood aspects of electrical work. This quiz will test your understanding of both NEC requirements and practical applications. The scenario questions are particularly important, as they reflect real-world situations where your knowledge could prevent hazards and ensure code compliance.

Multiple-Choice Questions

1. **Which of the following is NOT a required grounding electrode according to NEC 250.50 if present at the building?**

   a) Metal underground water pipe in direct contact with earth for at least 10 feet
   b) Metal gas piping system
   c) Concrete-encased electrode (Ufer groun
   d) Metal frame of the building that is effectively grounded

2. **According to the NEC, what is the minimum size copper grounding electrode conductor required for a service with 3/0 AWG copper phase conductors?**

   a) 8 AWG copper
   b) 6 AWG copper
   c) 4 AWG copper
   d) 2 AWG copper

3. **When are multiple disconnecting means permitted to be installed as service equipment?**

   a) Never; a single main disconnect is always required
   b) Only in residential occupancies
   c) Up to six disconnects are permitted
   d) Only when a main bonding jumper is installed at each disconnect

4. **What is the maximum size copper grounding electrode conductor required when connecting to a ground rod, regardless of service size?**

    a) 4 AWG
    b) 6 AWG
    c) 8 AWG
    d) 2 AWG

True/False Questions

5. **True or False:** A ground rod driven 6 feet into the soil meets NEC requirements as long as the top of the rod is flush with or below grade level.

6. **True or False:** When installing a subpanel in a detached building, the neutral and ground buses must be bonded together just as they are in the main service panel.

7. **True or False:** The equipment grounding conductor for a branch circuit can be smaller than the ungrounded (hot) conductors if the overcurrent device is 30 amperes or less.

8. **True or False:** A metal water pipe electrode must be supplemented by an additional electrode per NEC requirements.

Scenario-Based Questions

9. **Scenario**: You are installing a 200-amp service in a new residence. During excavation, you notice the concrete foundation was poured with reinforcing steel. The water service to the building is plastic. What grounding electrode system is required by the NEC in this situation?

10. **Scenario**: While renovating an older home, you learn that a bathroom circuit has a three-prong receptacle, but the box has no equipment grounding conductor. The wiring is old two-wire non-metallic cable. Describe your NEC-compliant options for this situation.

11. **Scenario**: You are installing a subpanel in a detached garage fed from the main house panel. Your feeder consists of two hot wires, a neutral, and an equipment grounding conductor in PVC conduit. Explain how you would configure the neutral and ground bars in this subpanel, citing the relevant NEC section.

12. **Scenario**: You're troubleshooting a residential service where electronic equipment frequently fails during thunderstorms. You discover the grounding electrode system consists of a single 8-foot ground rod with a measured resistance of 75 ohms to earth. What steps would you take to improve this grounding system and why?

13. **Scenario**: During a service upgrade at a commercial building, you find that the building's steel frame, concrete-encased electrode, and metal water piping are all present, but only the water pipe is connected to the grounding electrode system. Describe what corrections are needed and why.

14. **Scenario**: You're installing a 100A feeder to a workshop 200 feet from the main building. Due to voltage drop concerns, you've increased the feeder conductors from 3 AWG copper to 1 AWG copper. The overcurrent protection remains at 100A. What size equipment grounding conductor is required? Explain your reasoning and cite the relevant NEC section.

15. **Scenario**: On a jobsite, another electrician tells you that it's acceptable to use EMT conduit as the equipment grounding conductor for a 277V lighting circuit without pulling a separate ground wire. Evaluate this claim, explaining whether it's correct and under what conditions it might or might not be acceptable.

16. **Scenario**: You're troubleshooting a commercial kitchen where employees report getting shocks when touching the stainless steel sink and a nearby refrigerator simultaneously. Your voltage tester shows 3.5V between these two objects. What is likely causing this issue, and how would you resolve it?

17. **Scenario**: While inspecting a 480V motor installation in a facility with corrosive environmental conditions, you notice green corrosion on the equipment grounding conductor connection at the motor. The motor appears to be operating normally. Explain the potential hazard this presents and how it should be addressed.

18. **Scenario**: At a construction site with temporary power, workers report receiving mild shocks from a metal tool shed when it rains. The temporary power system includes a GFCI-protected panel, but the GFCIs don't trip. Based on your knowledge of grounding systems, explain what might be happening and how to correct it.

Answer Key and Explanations

1. **Answer: b) Metal gas piping system**

   Metal gas piping is specifically not permitted to be used as a grounding electrode per NEC 250.52(B)(1). The other listed items are all required electrodes if present at the building per 250.50.

2. **Answer: b) 6 AWG copper**

   According to Table 250.66, a service with 3/0 AWG copper phase conductors requires a 6 AWG copper grounding electrode conductor.

3. **Answer: c) Up to six disconnects are permitted**

   NEC 230.71 permits up to six disconnects for each service grouped in one location.

4. **Answer: b) 6 AWG copper**

   NEC 250.66(A) specifies that a grounding electrode conductor to a ground rod need not be larger than 6 AWG copper.

5. **Answer: False**

   NEC 250.53(G) requires that ground rods be driven to a depth of at least 8 feet. A 6-foot depth would not comply.

6. **Answer: False**

   In a separate building, the neutral and ground must be kept separate according to NEC 250.32(B) when an equipment grounding conductor is run with the feeder.

7. **Answer: True**

   According to Table 250.122, the equipment grounding conductor size is based on the rating of the overcurrent device and will typically be smaller than the ungrounded conductors for circuits 30A and below.

8. **Answer: True**

   NEC 250.53(D)(2) requires that a metal underground water pipe electrode be supplemented by an additional electrode.

9. **Answer**:

   The NEC requires that all available grounding electrodes be used together to form the grounding electrode system (250.50). In this case, the reinforcing steel in the concrete foundation likely qualifies as a concrete-encased electrode (Ufer ground) if it includes at least 20 feet of bare or electrically conductive coated steel reinforcing bars or rods of not less than ½" diameter, or 20 feet of bare copper conductor not smaller than 4 AWG encased in at least 2" of concrete (250.52(A)(3)). Since this electrode exists, it must be used. Additionally, at least one other electrode, typically a ground rod, must be installed to form a complete grounding electrode system since there is no metal water pipe electrode. Both electrodes must be bonded together with a properly sized grounding electrode conductor based on the service size (6 AWG copper for a 200A service with copper conductors).

10. **Answer**: The NEC provides several options for this situation under 406.4(D):

    * Replace the receptacle with a non-grounding type (two-prong) receptacle
    * Replace it with a GFCI-protected receptacle and mark it "No Equipment Ground.»
    * Replace it with a standard grounding receptacle protected by a GFCI at an upstream location and mark it "GFCI Protected" and "No Equipment Ground."
    * Install a new equipment grounding conductor from the receptacle to the service equipment, and use a grounding-type receptacle

    Options 2 and 3 provide shock protection without an equipment ground path. Option 4 provides both shock protection and equipment grounding but requires running a new wire, which may be difficult in finished walls.

11. **Answer**: In a detached building fed with a feeder that includes an equipment grounding conductor, NEC 250.32(B) requires the neutral and equipment grounding conductors to be kept separate. The neutral bar must be isolated from the panel enclosure (not bonded to it), while the equipment grounding conductor must be bonded to the enclosure. This prevents parallel current paths on the grounding and neutral conductors. Practically, this means installing an insulated neutral bar that doesn't contact the metal enclosure, and either using the panel's ground bar (which is bonded to the enclosure) or installing a separate ground bar that is bonded to the enclosure. All equipment grounding conductors connect to the ground bar, and all neutrals connect to the isolated neutral bar.

12. **Answer**: The high resistance reading of 75 ohms indicates poor ground contact, which explains the equipment failures during storms. NEC 250.53(A)(2) requires that if a single rod, pipe, or plate electrode has a resistance to ground over 25 ohms, it must be supplemented by an additional electrode. I would install a second ground rod at least 6 feet away from the first and connect them with a properly sized grounding electrode conductor. The NEC doesn't require resistance testing after installing the second rod. If the soil conditions are particularly poor, I might consider:

    * Installing chemical ground enhancement material around the rods
    * Using longer rods (e.g., 10 feet instead of 8 feet)
    * Installing a ground ring electrode around the building
    * Adding a concrete-encased electrode, if possible, during future construction

13. **Answer**: NEC 250.50 requires that all available grounding electrodes present at the building be bonded together to form the grounding electrode system. The building's steel frame and concrete-encased electrode are both required electrodes if they meet the definitions in 250.52(A), but they have not been connected to the grounding system. The correction needed is to run properly sized grounding electrode conductors to both the building steel and the concrete-encased electrode, and bond them to the existing grounding electrode

system. Per NEC 250.64(D), these conductors can be spliced using listed means such as irreversible compression connectors or exothermic welding. This creates a more effective grounding system with multiple paths to earth, providing better protection against lightning and fault currents.

14. **Answer**: When conductors are increased in size for voltage drop, NEC 250.122(B) requires that the equipment grounding conductor be increased proportionally. The calculation requires:

    * Determining the minimum EGC size from Table 250.122 for a 100A circuit (8 AWG copper)
    * Calculating the proportional increase based on the circular mil area of the conductors

15. Since 3 AWG has a circular mil area of 52,620 cmil and 1 AWG has 83,690 cmil, the increase ratio is 83,690 ÷ 52,620 = 1.59. The 8 AWG EGC has a circular mil area of 16,510 cmil. Increasing this proportionally: 16,510 × 1.59 = 26,251 cmil. The next standard size larger than 26,251 cmil is 6 AWG (26,240 cmil). Therefore, a 6 AWG copper EGC is required.

16. **Answer**: The electrician's claim can be correct under certain conditions. NEC 250.118 lists the types of equipment permitted as equipment grounding conductors, including Electrical Metallic Tubing (EMT). However, this is subject to several conditions:

- The conduit must be properly installed with appropriate fittings
- All connections must be properly tightened and secured
- The installation must maintain electrical continuity
- If installed in areas subject to severe physical damage or corrosive conditions, a separate equipment grounding conductor may be required.

17. **Answer:** For a standard 277V lighting circuit in a normal environment with properly installed EMT, the conduit itself is an acceptable equipment grounding conductor according to the NEC, and a separate ground wire is not required. However, many specifications for commercial and institutional buildings require a separate equipment grounding conductor regardless of the raceway type as an added safety measure.

18. **Answer**: The voltage difference between the sink and refrigerator indicates a bonding problem. NEC 250.104(A) requires metal water piping systems to be bonded to the service equipment enclosure, the grounded conductor at the service, or the grounding electrode conductor. The most likely cause is that the metal water piping system (including the stainless steel sink) is not properly bonded to the electrical grounding system. This creates a

difference in potential between the properly grounded refrigerator and the unbonded sink. To resolve this issue:

* Install a properly sized bonding jumper between the metal water piping system and the electrical service grounding system
* Ensure all connections are made with listed clamps or fittings appropriate for the application
* Verify that no plastic pipe sections have been installed that might interrupt the electrical continuity of the piping system
* Test for voltage differences after making the corrections

The size of the bonding jumper must comply with NEC 250.104(A), based on the size of the service-entrance conductors.

19. **Answer**: The green corrosion indicates a deteriorating connection, likely from galvanic action between dissimilar metals or environmental corrosion. While the motor appears to be operating normally, this presents a serious hazard because:

* The corroded connection increases resistance in the equipment grounding path
* During a fault, this increased resistance could prevent sufficient current flow to trip the overcurrent device quickly
* The high-resistance connection could cause the motor frame to become energized at a dangerous voltage during a fault
* The 480V system presents a higher danger than lower-voltage systems if a ground fault occurs

This should be addressed by:

* De-energizing the circuit
* Cleaning or replacing the corroded connection
* Using corrosion-resistant connectors appropriate for the environment
* Applying anti-oxidant compound to the connection
* Considering the use of sealed connections or regular maintenance inspections in corrosive environments
* Testing the ground path resistance after repairs

20. **Answer**: This scenario suggests a classic "neutral-ground voltage" situation commonly found at temporary sites. The likely causes and solutions are:

* The tool shed is probably grounded through its own ground rod or physical contact with the earth, creating a separate grounding system from the temporary power panel
* Under normal conditions, there's a small voltage potential between these two separate

grounding systems

* When it rains, the resistance of the earth decreases, allowing current to flow more easily between the two grounding points when workers touch both systems
* The GFCI doesn't trip because the current isn't flowing through the GFCI's sensing circuit (it's returning via the earth, not the GFCI's neutral)
* To correct this issue:
* Bond the tool shed to the temporary power system's grounding electrode system
* Run an equipment grounding conductor from the temporary panel to the tool shed
* Ensure a proper equipotential bonding system between all grounded metal objects
* Test for voltage differences after making these corrections

This creates a single grounding system rather than two competing systems, eliminating the potential difference that causes the shocks.

10

Electrical Safety and OSHA Standards

Six hundred times per second. That's how often your heart's electrical system pulses to keep you alive. And just once—just one brief contact with the wrong voltage—is all it takes to stop it forever.

I've watched seasoned journeymen with decades of experience make split-second decisions that ended careers. I've seen "routine" tasks turn catastrophic when safety protocols were treated as optional guidelines rather than life-preserving necessities. The most chilling part? In nearly every case, the electrician knew better. They simply believed it wouldn't happen to them.

Electricity doesn't care about your experience level. It doesn't respect your credentials or your confidence. It follows immutable physical laws that deal punishment with cold precision when we let our guard down. The 277-volt lighting circuit that "probably" isn't energized... The lockout procedure that seems "excessive" for a five-minute job... The arc flash boundary that feels "overly cautious." These are the moments where lives balance on the edge of disaster.

This chapter isn't about passing an exam or checking regulatory boxes. It's about ensuring you make it home each night with your body intact and your future secure. We'll explore OSHA requirements not as bureaucratic hurdles but as hard-won wisdom written in the blood and pain of those who came before us. We'll examine arc flash calculations not as abstract formulas but as mathematical shields against forces that can vaporize metal in milliseconds.

The most dangerous phrase in our profession isn't "high voltage"—it's "I've always done it this way." The knowledge in this chapter represents the difference between confidence and competence, between luck and preparation.

Because when it comes to electrical work, you rarely get a second chance to learn from your mistakes.

10.1 Lockout/Tagout (LOTO) Procedures

"Is it off?" Those three words might be the most important question in electrical work. Not "Is the switch in the off position?" or "Did someone say it was de-energized?" but actually, verifiably OFF—with absolute certainty that it cannot be re-energized while you're working.

Lockout/Tagout procedures form the foundation of electrical safety. These aren't just regulatory requirements—they're life-preserving protocols that stand between electricians and potentially fatal hazards. The process of isolating energy sources, verifying de-energization, and physically preventing re-energization has saved countless lives in our industry.

I once watched a contractor's employee begin working on what he thought was a locked-out circuit, only to discover—through painful electrical shock rather than proper testing—that he had locked out the wrong breaker. His mistake? Assuming that the panel schedule was accurate, rather than verifying with proper testing. His shortcut nearly cost him his life.

In this section, we'll examine the formal LOTO procedures required by OSHA, the specific steps that ensure energy control, and the practical application of these protocols in real-world electrical work. Whether you're working on small residential services or complex industrial systems, these procedures apply with equal importance.

10.2 The Importance of LOTO

Preventing Electrical Shock: By locking out power sources, electricians eliminate the risk of accidental shock when working on equipment.

Electricity seeks the path of least resistance to ground, and that path can easily become the human body. At household voltage (120V), the resulting shock can cause muscle contractions strong enough to prevent letting go of an energized conductor. At higher voltages common in commercial and industrial settings (277V, 480V), the current passing through the body can cause ventricular fibrillation, a chaotic heart rhythm incompatible with life.

Lockout procedures create a physical barrier between workers and these lethal hazards. By securing energy sources in the "off" position, LOTO eliminates the possibility of unexpected energization during maintenance or repair work.

A journeyman I worked with in my apprentice days had a permanent reminder of inadequate lockout; it was a jagged scar running from his palm to his elbow. He had been working on what he thought was a de-energized 277V lighting circuit when someone restored power without warning.

The momentary contact resulted in third-degree burns, nerve damage, and a three-month recovery. His experience taught me that proper LOTO isn't just about compliance—it's about survival.

Accident Prevention: LOTO procedures also help prevent the unintentional start-up of machinery or electrical systems during maintenance.

Electrical work often involves multiple power sources, complex control systems, and equipment that can move or actuate unexpectedly. Modern equipment may have automated functions, remote activation capabilities, or stored energy that can be released without warning.

LOTO protocols address these complexities by requiring systematic identification of all energy sources and methodical verification that each has been properly controlled. This comprehensive approach prevents the surprises that lead to accidents.

Consider a conveyor system with multiple motors, control circuits, and pneumatic components. Without proper LOTO, someone might disconnect the main power while forgetting about backup systems, control transformers, or stored pneumatic energy. Each overlooked energy source represents a potential hazard to workers.

OSHA Compliance: Adhering to LOTO procedures is required by OSHA, ensuring compliance with workplace safety regulations.

OSHA standard 29 CFR 1910.147 (Control of Hazardous Energy) and 1910.333 (Selection and Use of Work Practices) mandate specific LOTO requirements. These aren't optional guidelines—they're federal regulations with the force of law, developed in response to workplace fatalities and injuries.

OSHA citations for LOTO violations consistently rank among the most frequent across all industries. Penalties can range from several thousand dollars for minor violations to six figures for willful or repeated non-compliance. Beyond financial penalties, LOTO violations can result in stop-work orders, increased insurance costs, and even criminal prosecution in cases of serious injury or death.

More importantly, these regulations exist because properly implemented LOTO procedures work. When followed diligently, they prevent the accidents, injuries, and fatalities that occur when hazardous energy is not properly controlled.

10.3 Steps for Lockout/Tagout

The difference between a proper LOTO procedure and a shortcut can be the difference between life and death. Each step in this process serves a specific safety function and must be performed in sequence, without exception.

1. Preparation

Identify and understand all energy sources associated with the equipment to be worked on.

Before any lockout begins, you must thoroughly identify every energy source connected to the equipment—not just the main power supply. This includes:

- Primary electrical feeds
- Secondary or control power supplies
- Backup power sources
- Capacitors and other stored electrical energy
- Pneumatic or hydraulic systems
- Mechanical energy (springs, counterweights, elevated components)
- Thermal energy (hot surfaces, steam)

Many accidents occur because workers fail to recognize all energy sources. On complex equipment, this step may require reviewing schematics, consulting with operators, or tracing circuits to ensure nothing is overlooked.

2. Notification

Inform all affected employees of the shutdown and lockout procedure.

This communication step prevents confusion and ensures everyone knows that equipment will be unavailable. Notification should include:

- which equipment will be locked out
- why the lockout is necessary
- approximately how long the equipment will be unavailable
- who is performing the work
- who to contact with questions or concerns

In larger facilities, this might involve formal work orders, radio announcements, or shift briefings. In smaller operations, direct communication with affected workers may be sufficient.

3. Shutdown

De-energize the equipment by turning off all power supplies, ensuring no energy can be released.

Following normal shutdown procedures protects both the equipment and workers. An orderly shutdown:

- Prevents damage to equipment from sudden power loss
- Allows controlled dissipation of stored energy
- Prevents unexpected system responses
- Ensures all components reach a zero-energy state

Shutdown should follow the manufacturer's recommended sequence or the facility's established procedures, especially for complex systems with multiple components.

4. Lockout

Physically lock the power sources to prevent re-energization. Use locks and tags to indicate that work is in progress.

This critical step physically prevents the restoration of energy sources. Proper lockout requires:

- Each authorized worker applies their personal lock to each energy isolation device.
- Each lock has only one key, which remains with the worker.
- Tags clearly identify who placed the lock, when it was placed, and why.
- Multiple locks are accommodated using hasps or lockout boxes when necessary.
- Lockout devices are substantial enough to prevent removal without excessive force.

The fundamental principle here is that each person working on the equipment controls their own protection through their personal lock. No one else should ever be able to remove someone's lock while they're still working.

5. Verification

Test the equipment to verify that it is properly de-energized before beginning any work.

Verification confirms that the lockout was effective and all energy sources are controlled. This step must include:

- Attempting to operate the equipment (pressing start buttons, etc.)
- Testing all circuits with appropriate testing devices
- Confirming zero energy state of all components
- Checking for stored energy in capacitors, springs, or accumulators
- Verifying test equipment on a known source before and after testing

The verification step is not optional, even when you're "certain" the power is off. Testing must be performed as if your life depends on it—because it does.

6. Release from Lockout

After work is completed, follow the correct procedure to remove locks and restore the system to normal operation.

Proper release procedures ensure that equipment return to service happens safely:

- Inspect the work area to ensure all tools and materials are removed.
- Verify all workers are clear of the equipment.
- Each worker removes only their own lock and tag.
- Notify all affected employees before re-energizing.
- Methodically restore energy in the proper sequence.
- Monitor initial operation to ensure proper function.

The integrity of the LOTO process depends on strict adherence to these steps. Shortcuts, workarounds, or exceptions undermine the entire safety system and create potentially fatal risks.

10.4 Personal Protective Equipment (PPE) Requirements

"It's just a quick check—I don't need gloves for this."

Those words have preceded countless electrical accidents. I've watched experienced electricians reach into live 277V junction boxes wearing nothing but a short-sleeve shirt and a confident smile. I've seen journeymen test 480V gear without face shields because "it's too hot to wear all that." And I've visited colleagues in burn units who would give anything for a second chance to put on the PPE they deemed "unnecessary."

Personal Protective Equipment represents your last line of defense when something goes catastrophically wrong. While proper planning, de-energizing, and lockout procedures should be your primary protection, PPE becomes the critical barrier between you and devastating injury when those safeguards fail or when working on energized equipment becomes unavoidable.

The burn threshold for human skin is just 1.2 calories per square centimeter—about the energy in a small sip of hot coffee. Yet an arc flash from a typical 480V panel can deliver 40 calories/ cm^2 or more to exposed skin. Those numbers explain why proper PPE isn't a convenience or a recommendation—it's a life-preserving necessity that deserves unwavering commitment.

In this section, we'll explore the specific PPE requirements for electrical work, how to select appropriate protection based on hazard analysis, and the standards that govern these critical safety components.

10.5 What Is PPE?

Personal protective equipment (PPE) refers to the clothing and devices worn to protect electricians from electrical hazards, including shock, arc flash, and burns. The correct use of PPE is essential to minimize the risk of injury on the job.

PPE serves as the final barrier between an electrician and potentially life-altering injuries. While proper work practices and engineering controls should be the primary methods of protection, PPE becomes crucial when these measures aren't sufficient or when energized work is unavoidable. Properly selected and worn PPE can mean the difference between a minor incident and a career-ending—or life-ending—injury.

Key Types of PPE for Electricians

Insulated Gloves: Protect against electrical shock, especially when working on live circuits.

Electrical insulating gloves are rated by class, with each class protecting different voltage levels:

- Class 00: Maximum use voltage of 500V AC
- Class 0: Maximum use voltage of 1,000V AC
- Class 1: Maximum use voltage of 7,500V AC
- Class 2: Maximum use voltage of 17,000V AC
- Class 3: Maximum use voltage of 26,500V AC
- Class 4: Maximum use voltage of 36,000V AC

These rubber insulating gloves must be air-tested before each use to check for punctures or tears, and they require regular electrical testing (typically every 6 months). Leather protector gloves are worn over the rubber insulators to prevent physical damage.

Arc Flash Suits: Specialized suits that provide protection against burns and flying debris during an arc flash event.

Arc flash suits are rated in calories per square centimeter (cal/cm^2), indicating the level of thermal energy they can withstand without breaking open or igniting. They typically include:

- Coveralls or jacket/pants combinations made of arc-rated materials
- Arc-rated hood with face shield
- Arc-rated gloves
- Insulating footwear

The level of protection required depends on the incident energy calculation for the specific equipment being worked on, which is determined through an arc flash hazard analysis.

Hard Hats: Protect against falling objects and head injuries.

For electrical work, hard hats must be:

- Class E (Electrical) rated, tested to withstand 20,000 volts
- Free from conductive attachments or modifications
- Regularly inspected for cracks, dents, or degradation
- Replaced according to manufacturer guidelines (typically every 5 years)

Safety Glasses or Goggles: Shield eyes from flying debris or sparks.

Eye protection for electrical work should:

- Meet ANSI Z87.1 standards with side shields
- Be non-conductive
- Provide appropriate visible light filtration when working with intense light sources
- Be compatible with face shields and other PPE components

For higher hazard levels, arc-rated face shields or hood shields provide additional protection against arc flash energy.

Flame-Resistant (FR) Clothing: Specialized clothing to protect against burns from arc flashes.

FR clothing is designed to resist ignition and self-extinguish if it does catch fire. Key features include:

- Arc rating measured in cal/cm^2
- NFPA 70E hazard risk category (HRC) level designation
- Durability through multiple launderings
- Appropriate coverage of all exposed skin

FR clothing ranges from simple long-sleeve shirts and pants for minimal hazards to multi-layer suits for high-energy environments. Unlike regular clothing that can melt onto skin during an arc flash, FR materials provide critical seconds of protection that can prevent third-degree burns.

Hearing Protection: Used in environments with high noise levels from machinery or equipment.

While not specifically for electrical hazards, hearing protection is often necessary in electrical work environments due to:

- Generator noise
- Transformer hum
- Motor operation
- General industrial noise

Types include expandable foam earplugs, pre-molded reusable plugs, and earmuffs, each with different Noise Reduction Ratings (NRR).

Selecting the appropriate PPE requires understanding the specific hazards present, the task being performed, and the protection level required. NFPA 70E provides detailed tables and guidelines for selecting PPE based on task and equipment type, making it an essential reference for all electrical workers.

10.6 OSHA PPE Standards

A safety mill manager once handed me a written notice when he'd observed that I wasn't wearing any PPE. When I explained that I was just doing a quick inspection and not touching anything, he slid a photograph across the desk showing me leaning into an open 480V panel. Next to it, he placed a second photo—an employee from another plant with third-degree burns across his face from an arc flash that occurred during a similar "quick inspection."

"OSHA standards aren't suggestions," he said quietly. "They're written in the blood and pain of people who thought the rules didn't apply to them."

That lesson has stayed with me for decades. OSHA standards represent more than regulatory compliance—they establish the minimum requirements for workplace safety based on hard-learned lessons from industrial accidents. These standards don't exist to make work more difficult; they exist to ensure workers return home safely at the end of each day.

Choosing the Right PPE

Selecting appropriate PPE isn't a matter of grabbing whatever gear is available or following a one-size-fits-all approach. It requires careful analysis of specific hazards and understanding the protection levels different equipment provides. The wrong PPE can create a dangerous false sense of security while offering inadequate protection against actual hazards.

Assessment of Hazards

Identify the specific electrical hazards present in the workplace.

This assessment must be both systematic and comprehensive, considering:

Shock hazards: Evaluate voltage levels, equipment condition, and work tasks to determine shock risks. A worker replacing a 277V ballast faces different shock hazards than someone maintaining a 4160V motor starter.

Arc flash potential: Calculate incident energy levels at various working distances. These calculations consider available fault current, clearing times of protective devices, and working distance to determine the thermal energy workers might face during an arc flash event. Most facilities now require a formal arc flash study that labels equipment with specific hazard information.

Arc blast force: Beyond thermal energy, arc flashes create pressure waves and propel molten metal. These physical forces require consideration when selecting face and body protection.

Environmental factors: Evaluate temperature, confined spaces, and other conditions that might affect PPE effectiveness or create additional hazards.

The hazard assessment must be documented and updated whenever equipment or work processes change. NFPA 70E provides guidelines for performing these assessments, including tables that match common tasks with appropriate PPE categories.

Selecting Appropriate Gear

Choose PPE that offers protection for identified hazards based on risk assessment.

Proper selection requires understanding the protection ratings of various PPE components:

For shock protection: Select insulated gloves with appropriate voltage ratings (Class 00 through Class 4) based on the maximum voltage present. Remember that "maximum voltage" means phase-to-phase voltage, not phase-to-ground.

For arc flash protection: Match clothing and equipment ratings (in cal/cm^2) to the calculated incident energy levels. NFPA 70E defines PPE categories:

- Category 1: Minimum 4 cal/cm^2 (typically FR shirt and pants)
- Category 2: Minimum 8 cal/cm^2 (typically FR shirt, pants, and balaclava)
- Category 3: Minimum 25 cal/cm^2 (typically two-layer FR system)
- Category 4: Minimum 40 cal/cm^2 (typically multi-layer systems with hood and gloves)

For specific body parts: Select task-appropriate head, face, hand, and foot protection that integrates with your overall PPE system. All components must work together without compromising protection.

Always select PPE that meets relevant standards (ASTM, ANSI) and is properly sized for the individual worker. Ill-fitting PPE can reduce protection levels and create additional hazards.

Regular Inspections

PPE must be inspected regularly for wear and tear, and it should be replaced as needed.

Each type of PPE requires specific inspection procedures:

Insulating gloves: Before each use, perform visual inspection and air tests to check for punctures, tears, or chemical damage. Formal electrical testing is required every six months for in-service gloves.

Arc-rated clothing: Inspect for holes, tears, contamination with flammable substances, and reduced protection from multiple launderings. Most FR clothing has a specified lifetime in terms of wash cycles.

Face shields and eye protection: Check for scratches, cracks, or distortion that could compromise protection or visibility. Even minor damage can significantly reduce the protective value.

Hard hats: Examine for cracks, dents, or signs of UV degradation (fading, chalking). Test suspension systems for elasticity and proper function.

Documenting these inspections creates accountability and helps track PPE lifecycle. Many companies implement formal inspection schedules and documentation procedures, especially for high-voltage insulating equipment.

The right PPE, properly inspected and maintained, provides essential protection when other safety measures fail. However, PPE should always be considered the last line of defense after proper work planning, de-energizing procedures, and other administrative controls have been implemented.

OSHA also mandates that employers provide the necessary PPE and ensure its proper use on the job site. They also require regular training on how to use and maintain PPE correctly.

10.7 Workplace Hazards and How to Mitigate Them

Every electrical workplace contains hidden dangers. From exposed live parts to inadequate working clearances, from improper grounding to unexpected stored energy—these hazards are part of our daily environment as electricians. Recognizing them isn't just about compliance or avoiding citations; it's about identifying real threats before they cause harm.

The most effective electricians develop a sixth sense for spotting potential problems. They scan work areas systematically, question assumptions about what's energized, and constantly update their mental risk assessment as conditions change. This awareness doesn't come naturally—it's developed through experience, training, and sometimes painful lessons.

In this section, we'll examine the most common workplace hazards found in electrical environments and practical approaches to eliminate or control them. Rather than focusing on abstract regulations, we'll look at real-world situations and proven solutions that experienced electricians use to keep themselves and their colleagues safe.

10.8 OSHA Standards for Electrical Safety

OSHA's electrical safety standards directly impact how you perform your work as an electrician. Let's examine the specific requirements you need to understand and implement daily:

Electrical Safety-Related Work Practices

29 CFR 1910.333—Selection and Use of Work Practices

This standard requires that:

- You must de-energize equipment before working on it, except in very limited circumstances.
- Live parts operating at 50V or more must be de-energized or guarded by approved barriers.
- When working near overhead lines, OSHA mandates specific clearances: for lines up to 50kV, stay at least 10 feet away; for lines over 50kV, add 4 inches for each 10kV over 50kV.
- Portable ladders used around electrical work must have non-conductive side rails.
- Conductive items like tape measures and metal jewelry must not be worn when working near exposed energized parts.

29 CFR 1910.334 —Use of Equipment

This section specifies that:

- Portable electric equipment must be handled in a way that won't damage cords or equipment.
- Power and lighting circuits must be disconnected before lamp or fuse replacement.
- Reclosing circuits after protective device operation is prohibited until the cause is determined.
- Test instruments must be verified on a known source before and after testing to ensure proper operation

29 CFR 1910.335 —Safeguards for Personnel Protection

This details PPE requirements:

- Insulating protective equipment must be appropriate for the parts being worked on.
- When working near exposed energized conductors, you must use insulated tools.
- Protective shields, barriers, or insulating materials must be used when working near exposed energized parts.
- Each employee working on electrical systems must be provided with appropriate PPE based on the hazards present.

Hazardous Location Requirements

Class I Locations (29 CFR 1910.307)

For areas with flammable gases or vapors:

- Division 1 requires explosion-proof equipment, hermetically sealed equipment, or purged/pressurized systems.
- All threaded conduits must engage at least 5 full threads.
- Seals must be installed at boundaries between classified and unclassified areas.
- Specific wiring methods like threaded rigid metal conduit or Type MI cable are required.

Class II Locations

For combustible dust environments:

- All enclosures must be dust-tight.
- Motors must have dust-ignition-proof enclosures with maximum surface. temperatures below the ignition temperature of the dust.
- Junction boxes must have at least 5 inches of free space from any surface where dust might collect.

Class III Locations

For areas with ignitable fibers/flyings:

- All equipment must prevent the entrance of fibers/flyings.
- Surface temperatures must stay below 165°C (329°F).
- Lighting fixtures must have suitable guards to prevent contact with the lamp.

OSHA enforces these standards through workplace inspections, and violations can result in citations with penalties ranging from $15,625 for serious violations to $156,259 for willful or repeated violations (as of 2023).

These standards are implemented alongside the National Electrical Code (NEC/NFPA 70), which OSHA incorporates by reference. While the NEC focuses on proper installation, OSHA standards emphasize safe work practices around electrical equipment, creating complementary layers of protection.

10.9 Mitigating Workplace Hazards

Proper Training

Regular safety training ensures that workers understand how to recognize and avoid hazards.

Effective electrical safety training must include specific, practical skills:

Voltage Testing Procedure Training: Workers need to learn the specific sequence:

1. Verify tester operation on a known live source.
2. Test the circuit in question.
3. Verify tester again on a known source.

Many electrical accidents happen because workers skip steps in this sequence, leading to false "de-energized" readings from faulty test equipment.

Arc Flash Boundary Training: Workers must know exactly what these boundaries mean:

- Limited Approach Boundary (typically 42 inches for 480V equipment): Where unqualified persons must stop.
- Restricted Approach Boundary (typically 12 inches for 480V): Where shock protection techniques are required.
- Arc Flash Boundary (often 10+ feet): Where PPE is required to prevent burn injury.

Knowing these specific distances for the equipment you work with can prevent serious injuries.

Emergency Response Training: Training must include specific actions for electrical accidents:

1. How to safely separate a shock victim from the circuit without becoming a victim yourself.
2. How to quickly shut down power in emergencies.
3. How to properly apply CPR when needed (renewal typically required every 2 years).

Equipment Maintenance

Keeping electrical tools and systems in good working condition requires specific, scheduled activities:

Insulation Testing: For power cables, perform megohmmeter testing:

- For low-voltage systems (<600V): Minimum acceptable reading is 1 megohm
- For medium voltage: Apply 2500V test voltage and look for readings of 5+ megohms

Readings declining over time indicate insulation deterioration requiring immediate attention.

Circuit Breaker Maintenance: Each type requires specific maintenance:

- Molded case breakers: Perform trip testing annually with specialized test equipment
- Air circuit breakers: Inspect arc chutes, clean mechanism, lubricate moving parts, and test protective functions every 1-3 years
- Vacuum breakers: Check contact wear indicators and perform contact resistance tests every 2-5 years

GFCI Testing: These critical safety devices must be tested:

- Monthly for portable GFCI devices
- Test by pressing the "TEST" button and verifying power disconnection
- Units that don't trip when tested must be immediately replaced, not just reset

Worksite Inspections

Regular checks of the worksite require specific focus areas:

Panel and Equipment Inspections: Look for these specific issues:

- Knockouts missing from panels (creates shock and arc flash hazards).
- Breakers that feel hot to the touch (indicates overloading or connection problems).
- Discoloration of terminals (indicates overheating from loose connections).
- Proper working clearance (minimum 3 feet in front of 0-150V equipment, 3.5-4 feet for 151-600V equipment).

Extension Cord Inspections: Check specifically for:

- Damaged outer jacket (immediate removal from service required)
- Missing ground pins (remove from service immediately)
- Proper sizing for the connected load (12 AWG minimum for construction sites)
- Placement away from walkways and protected from physical damage

Thermal Scanning —Conduct infrared inspections of electrical equipment:

- Connections showing 25°F above ambient require monitoring
- Connections 50°F above ambient need attention at next scheduled outage
- Connections 70°F+ above ambient require immediate repair

These specific inspection points and maintenance actions provide concrete steps to identify and address electrical hazards before they cause accidents. By implementing these precise practices, you create layers of protection that significantly reduce workplace electrical hazards.

10.10 Common Electrical Safety Violations

Every OSHA inspector has stories that make them shake their head in disbelief. The extension cord runs through standing water. The panel with every knockout removed and live bus bars exposed. The "temporarily" bypassed safety switch that's been that way for three years. These violations appear so frequently that inspectors can often predict what they'll find before they even enter a jobsite.

What's most disturbing isn't just the frequency of these violations—it's how normalized they become in workplace culture. Practices that violate safety standards often become "standard operating procedure" until an accident occurs. The electrician who tests with his finger rather than a meter. The contractor who treats lockout procedures as optional when "we're just going to be a minute." The supervisor who pressures workers to restore power before safety checks are complete.

Understanding the most common electrical safety violations helps you recognize these dangerous practices before they lead to incidents. More importantly, it helps you avoid the gradual drift toward accepting these violations as normal—a drift that has preceded countless preventable accidents.

Violation 1: Failure to De-Energize Equipment

Problem: One of the most serious violations is working on live electrical equipment without following proper lockout/tagout (LOTO) procedures.

Despite clear OSHA requirements, working on energized equipment continues to be one of the most frequent and dangerous violations in electrical work. Under 29 CFR 1910.333(a)(1), OSHA explicitly states: "Live parts to which an employee may be exposed shall be de-energized before the employee works on or near them, unless the employer can demonstrate that de-energizing introduces additional or increased hazards or is infeasible due to equipment design or operational limitations."

This requirement leaves little room for interpretation, yet violations occur daily across job sites of all sizes, from residential work to industrial facilities. Common scenarios include:

- "Quick" troubleshooting without proper PPE.
- "Just checking" voltage or current without following safety procedures.
- Failing to verify de-energization with proper testing.
- Working inside the minimum approach boundary without appropriate protection.
- Removing panel covers or opening equipment without shutting off power.

The justifications are always similar: "It would take too long to shut it down," "We can't interrupt the client's operations," or "I've done this hundreds of times without getting hurt." These rationalizations ignore the fundamental risk—electricity needs only a momentary lapse in concentration or a slight misstep to cause severe injury or death.

OSHA Penalty: This can result in severe fines and, more importantly, potential injury or death.

OSHA classifies this as a serious violation at minimum, with penalties typically starting at $15,625 per violation (as of 2023). In cases where the violation is deemed willful or repeated, penalties can reach $156,259 per violation.

Beyond financial penalties, the human cost is far more significant. Failure to de-energize equipment before working on it is a leading cause of electrical fatalities and serious injuries. These incidents often occur to experienced electricians who have become complacent about the dangers of energized work.

The most effective prevention is strict adherence to a "de-energize first" policy with very limited exceptions. When live work is truly necessary, it should occur only with:

- Written energized work permits.
- Detailed job hazard analysis.
- Appropriate PPE based on arc flash and shock hazard calculations.
- Qualified personnel who are specifically trained for energized work.
- Clear justification for why the circuit cannot be de-energized.

Remember that OSHA allows live work only in very specific circumstances—not for convenience, schedule pressure, or to avoid production disruptions.

Violation 2: Inadequate PPE Use

Problem: Not wearing the correct PPE or wearing worn-out equipment can lead to serious injuries, such as burns from arc flashes.

OSHA standards 29 CFR 1910.335 and 1926.416 specifically require appropriate protective equipment when working where potential electrical hazards exist. Despite these clear requirements, PPE violations are still some of the most frequently cited in the electrical trade.

These violations typically take several forms:

Completely Missing PPE: Electricians often perform "quick checks" or minor adjustments with no protective equipment at all. This commonly occurs when:

- testing a circuit they believe is de-energized
- resetting a tripped breaker
- making "minor" adjustments to energized equipment
- working on lighting circuits or control systems perceived as "low risk"

Insufficient PPE Level: Workers use some protection, but not enough for the hazard present. Examples include:

- Using standard work gloves instead of properly rated electrical insulating gloves.
- Wearing non-FR clothing when working on equipment with arc flash potential.
- Using basic safety glasses instead of full-face protection.
- Choosing lower-rated PPE than required by the arc flash hazard analysis.

Damaged or Expired PPE: Using protective equipment that no longer provides adequate protection.

- Electrical gloves with pinhole punctures or tears
- Arc-rated clothing that has been washed with fabric softeners or contaminated with flammable substances
- Face shields with cracks or excessive scratching
- Insulated tools with damaged insulation

Improperly Worn PPE: Using the right equipment but in the wrong way:

- Wearing rubber insulating gloves without leather protectors.
- Rolling up sleeves of FR clothing.
- Failing to secure face shields properly.
- Leaving gaps between protective layers.

Behind most of these violations lies a common thread: underestimation of the risk. An arc flash can occur in equipment that has operated normally for decades. Insulation that has functioned perfectly for years can suddenly fail. These hazards don't announce themselves in advance—which is precisely why PPE must be used consistently and correctly.

OSHA Penalty: Fines for failure to provide or use adequate PPE are common and significant. OSHA classifies these as serious violations because inadequate PPE directly exposes workers to hazards that can cause severe injury or death.

OSHA typically classifies PPE violations as "serious" because they present substantial probability of death or serious physical harm. Base penalties start at $15,625 per violation (as of 2023). When multiple workers are observed without proper PPE, each instance can be cited separately, quickly multiplying the financial impact.

Prevention requires both proper equipment selection and consistent enforcement:

- Perform hazard assessments to determine appropriate PPE for each task
- Provide workers with properly sized, high-quality protective equipment
- Implement regular inspection procedures for all PPE
- Remove damaged equipment from service immediately
- Create a culture where bypassing PPE requirements is never acceptable

Remember: PPE is your last line of defense when other protective measures fail. It must be selected, maintained, and used with the assumption that you will need its protection.

Isolation 3: Improper Grounding

Problem: Not grounding electrical systems correctly can result in electrical shocks, fires, and equipment failure.

OSHA standards 29 CFR 1910.304 (for general industry) and 1926.404 (for construction) specify detailed requirements for grounding of electrical systems and equipment. Despite these clear regulations, improper grounding remains a frequent citation.

These violations typically manifest in several ways:

Missing Equipment Grounding Conductors: This creates a situation where metal equipment enclosures are energized without a path to trip protective devices:

- Extension cords with the ground pin removed or ground wire disconnected.
- Power tools with damaged or modified grounding connections.
- Fixed equipment connected without proper equipment grounding.
- Temporary wiring installed without grounding conductors.

Improper Neutral-Ground Bonding

Creating multiple neutral-ground bonds in a system allows current to flow on equipment grounding conductors and metal parts:

- Connecting neutral and ground at subpanels (instead of keeping them separate).
- Installing a main bonding jumper at both the service equipment and downstream panels.
- Improper neutral-ground connections in temporary power setups or generators.
- Using the same conductor for both grounded and grounding functions.

Inadequate Grounding Electrode Systems

Failing to properly connect to earth:

- Using only a single ground rod without verifying resistance to earth.
- Failing to connect to available grounding electrodes, like building steel or water pipes.
- Undersized grounding electrode conductors.
- Poor connections to grounding electrodes that corrode over time.

Broken Ground Paths

Interruptions in the grounding system, like:

- Painted surfaces that prevent proper ground contact in enclosures.
- Loose conduit connections breaking equipment grounding continuity.
- Using non-metallic fittings in metallic raceway systems.
- Missing bonding jumpers across equipment like water meters or flexible connections.

What makes grounding violations so dangerous is that the system often appears to be functioning normally, despite these defects. Power flows, equipment operates, and no immediate symptoms indicate a problem. The hazard remains hidden until a fault occurs—at which point the lack of proper grounding can turn a manageable fault into a deadly situation.

OSHA Penalty: OSHA typically classifies grounding violations as "serious" because they create substantial risk of death or serious injury. Base penalties start at approximately $15,625 per violation (as of 2023). In cases of repeat violations or willful disregard for standards, penalties can increase substantially.

Beyond monetary penalties, OSHA has the authority to issue stop-use orders for equipment with serious grounding deficiencies. This can shut down critical production equipment until proper grounding is restored—often resulting in significant operational costs beyond the direct penalties.

Effective prevention of grounding violations includes:

- Regular testing of grounding continuity in equipment and systems.
- Comprehensive inspection of grounding connections during installation.
- Proper training on the importance of grounding and how to verify its integrity.
- Immediate correction of any identified grounding issues, regardless of operational impact.

Remember that proper grounding isn't just about code compliance—it's a fundamental safety system that protects workers from potentially fatal exposure to energized parts during fault conditions.

10.11 Quiz-Testing Your Knowledge of Electrical Safety and OSHA Standards

Put your electrical safety knowledge to the test with this quiz. These principles could save your life or prevent serious injury on the job. Don't just memorize answers—think through the practical application of each safety concept.

Multiple-Choice Questions

1. **According to OSHA, when is it acceptable to work on energized electrical equipment?**

    a) When it would be inconvenient to shut down the equipment
    b) When de-energizing would create additional hazards or is infeasible due to equipment design
    c) When the work will take less than 10 minutes
    d) When working in a residential setting

2. **What is the minimum approach boundary for unqualified persons to exposed energized parts operating at 480 volts?**

    a) 3 feet
    b) 3.5 feet
    c) 10 feet
    d) 42 inches

3. **Which of the following is NOT a required element of a proper lockout/tagout procedure?**

   a) Notification of affected employees
   b) Verification of de-energization
   c) Supervisor's signature on work permit
   d) Application of lockout devices

4. **What class of insulating gloves is required for work on a 480V system?**

   a) Class 00
   b) Class 0
   c) Class 1
   d) Class 2

5. **How often must rubber insulating gloves be electrically tested according to OSHA requirements?**

   a) Every month
   b) Every 3 months
   c) Every 6 months
   d) Every 12 months

True/False Questions

6. **True or False:** If an electrical panel is marked with an arc flash warning label indicating PPE Category 2, a worker must wear Category 2 PPE even when only operating a circuit breaker with the panel door closed.

7. **True or False:** OSHA allows qualified workers to remove the lock of another worker who has left the site if production requires it.

8. **True or False:** Extension cords can be used as a substitute for fixed wiring as long as they are in good condition and the proper gauge for the load.

9. **True or False:** When testing to verify a circuit is de-energized, you must test the voltage tester on a known live source before and after testing the circuit in question.

10. **True or False:** GFCI protection is required for all 125-volt, single-phase, 15—and 20-ampere receptacles used on construction sites.

Scenario-Based Questions

11. **Scenario: You arrive at a jobsite and are asked to replace a ballast in a fluorescent light fixture. The building manager says the circuit breaker has been turned off, but you don't know which one it is.**

    What is the proper procedure before beginning work?

12. **Scenario: While working on a control panel, you notice that the arc flash hazard label indicates an incident energy of 12 cal/cm² at the working distance.**

    What specific PPE would be required for this task if it must be performed while the panel is energized?

13. **Scenario: You observe a coworker using a metal ladder near exposed 277V lighting circuits. The worker is not making contact with the circuits but is within 3 feet of them.**

    What OSHA violation is occurring, and what is the correct action to take?

14. **Scenario: You need to diagnose an intermittent problem in a 480V motor starter that only occurs when the equipment is running. Your supervisor says you can just "quickly open the cover and look inside while it's running."**

    How should you respond to this request?

15. **Scenario: You are preparing to work on electrical equipment that has multiple energy sources: 480V electrical, pneumatic pressure, and hydraulic pressure.**

    Describe the complete lockout/tagout procedure you would follow before beginning work.

Answer Key and Explanations

1. **Answer: b) When de-energizing would create additional hazards or is infeasible due to equipment design**

   OSHA 29 CFR 1910.333(a)(1) requires circuits to be de-energized before work begins unless the employer can demonstrate that de-energizing introduces additional hazards or is infeasible due to equipment design or operational limitations. Convenience or time constraints are never acceptable reasons to work on energized equipment.

2. **Answer: d) 42 inches**

   According to NFPA 70E and OSHA interpretations, the limited approach boundary for unqualified persons at 480 volts is 42 inches (3.5 feet). This is the minimum distance unqualified persons must maintain from exposed energized parts.

3. **Answer: c) Supervisor's signature on work permit**

   While many companies require supervisor approval on energized work permits, OSHA's lockout/tagout standard (29 CFR 1910.147) does not specifically require supervisor signatures. The other options (notification, verification, and application of lockout devices) are all required elements of a proper LOTO procedure.

4. **Answer: b) Class 0**

   Class 0 gloves are rated for maximum use voltage of 1,000V AC and are appropriate for work on 480V systems. Class 00 gloves are rated only up to 500V AC, which would be insufficient for 480V work.

5. **Answer: c) Every 6 months**

   OSHA 29 CFR 1910.137 requires rubber insulating gloves to be electrically tested before first issue and every 6 months thereafter when in service. This testing must be performed according to ASTM standards by a qualified testing facility.

6. **Answer: False**

   Operating a circuit breaker with the panel door closed (normal operation) typically does not require the full PPE indicated on arc flash labels. NFPA 70E provides specific tables for common tasks like operating circuit breakers with equipment in normal condition. The arc flash label indicates PPE requirements for work inside the panel with covers removed.

7. **Answer: False**

   OSHA 29 CFR 1910.147(e)(3) explicitly prohibits one employee from removing another employee's lockout device. Each lock should only be removed by the person who applied it. Emergency lock removal procedures require specific protocols, documentation, and verification that the authorized employee is not at the facility.

8. **Answer: False**

   OSHA 29 CFR 1910.305(g)(1)(iii) specifically prohibits using flexible cords and cables as a substitute for fixed wiring. Extension cords are designed for temporary use with portable equipment, not as permanent installations, regardless of their condition or capacity.

9. **Answer: True**

   OSHA 29 CFR 1910.333(b)(2)(iv)(B) requires that test instruments be verified on a known voltage source before and after testing for the absence of voltage. This ensures that the tester was functioning properly at the time the circuit was tested.

10. **Answer: True**

OSHA 29 CFR 1926.404(b)(1)(ii) requires all 125-volt, single-phase, 15—and 20-ampere receptacles used on construction sites that are not part of the permanent wiring to have GFCI protection for personnel.

11. **Answer:** The proper procedure includes:

a) Identify the specific circuit controlling the light fixture
b) Apply your lock and tag to the correct circuit breaker
c) Verify lockout by attempting to restore power to the circuit
d) Test the actual fixture with a properly functioning voltage tester before beginning work:

i. First, test the voltage tester on a known live circuit
ii. Then test the locked-out circuit to verify it's de-energized
iii. Test the voltage tester again on a known live circuit to confirm it's still working

e) Only then begin work on the fixture

Never rely solely on someone else's assurance that power has been turned off. Always verify personally, following all steps of proper LOTO procedures.

12. **Answer:** For work on equipment with 12 cal/cm^2 incident energy, the required PPE includes:

a) Arc-rated clothing with a minimum rating of 12 cal/cm^2 (typically Arc-rated long-sleeve shirt and pants or coveralls)
b) Arc-rated face shield and balaclava or arc flash suit hood with minimum 12 cal/cm^2 rating
c) Arc-rated gloves
d) Hearing protection (ear canal inserts)
e) Heavy-duty leather footwear
f) No exposed skin on the face, neck, or hands

This corresponds to PPE Category 3 in the NFPA 70E-2021 standard. The selected PPE must always meet or exceed the calculated incident energy at the working distance.

13. **Answer:** The violation is using a metal ladder near exposed energized parts, which violates OSHA 29 CFR 1910.333(c)(7) requiring ladders with non-conductive side rails when working near exposed energized parts. The correct action is:

    a) Stop work immediately and inform the coworker of the danger
    b) Replace the metal ladder with a fiberglass or other non-conductive ladder
    c) Ensure proper approach boundaries are maintained based on the voltage level
    d) If the work requires proximity to exposed energized parts, implement appropriate PPE and work practices for electrical safety

    Metal ladders conduct electricity and create an unnecessary shock hazard when used near electrical equipment.

14. **Answer:** The proper response would be:

    a) Refuse to open the enclosure while energized without proper assessment and protection
    b) Explain that OSHA requires a justified reason for working on energized equipment
    c) Request a formal energized electrical work permit that includes:

        i. Justification for why the work cannot be performed de-energized
        ii. Shock and arc flash hazard analysis results
        iii. Required PPE based on the hazard analysis
        iv. Safe work procedures for the energized diagnosis

    d) Ensure you have the required training and PPE before proceeding
    e) Suggest alternative diagnostic methods that might not require exposure to energized parts

    "Quickly looking inside" energized 480V equipment violates OSHA standards and creates serious arc flash and shock hazards.

15. **Answer:** A complete LOTO procedure for equipment with multiple energy sources would include:

    a) Notify all affected employees about the shutdown
    b) Properly shut down the equipment following normal stopping procedures
    c) Isolate all energy sources:

        i. Turn off and lock out the electrical disconnect
        ii. Close and lock out pneumatic valves
        iii. Close and lock out hydraulic valves

    d) Dissipate stored energy:

        i. Release pneumatic pressure from lines and tanks
        ii. Bleed hydraulic pressure from the system

iii. Discharge any capacitors in the electrical system
iv. Block or secure any mechanical parts that could move

e) Verify zero energy state for each energy source:

i. Test for absence of voltage with a properly verified tester
ii. Confirm pneumatic pressure gauges read zero
iii. Verify hydraulic pressure gauges read zero
iv. Attempt to operate equipment to ensure it cannot start

f) Maintain lockout/tagout throughout the service
g) Verify safe conditions before restoring energy after work is complete

Multiple energy sources require controlling each source independently, with specific verification for each type of energy.

Assessment

Count your correct answers:

- 13-15 correct: Excellent understanding of electrical safety principles
- 10-12 correct: Good knowledge, but review the areas you missed
- 7-9 correct: Basic understanding with significant knowledge gaps
- Below 7: Serious review needed before working with electrical systems

Remember that electrical safety knowledge isn't just for passing tests—it's essential protection against very real hazards that can cause serious injury or death. Always prioritize safety over productivity or convenience.

11

Testing and Troubleshooting Equipment

The difference between guessing and knowing is measurement.

As electricians, we work with an invisible force. We can't see electricity flow through conductors, watch electrons jump across connections, or visually identify voltage differences. Instead, we rely on specialized testing equipment to make the invisible visible—transforming abstract electrical properties into concrete, measurable values that guide our decisions.

The best electricians aren't those who memorize the most codes or install the neatest panels. They're the ones who can walk onto a job where nothing works, methodically test the system, and zero in on the exact point of failure while others are still scratching their heads. This troubleshooting skill separates master craftsmen from parts-changers.

Testing equipment serves as our eyes and ears in the electrical world. Multimeters reveal whether voltage is present and at what magnitude. Ammeters tell us how much current is flowing without breaking the circuit. Megohmmeters expose insulation weaknesses before they become catastrophic failures. Each instrument provides a piece of the puzzle, helping us build an accurate picture of what's happening within the system.

In this chapter, we'll examine the essential testing equipment every journeyman electrician should know how to use, the proper techniques for accurate measurement, and systematic approaches to troubleshooting that transform random guesswork into logical problem-solving. These skills will serve you throughout your career, whether you're verifying a newly installed circuit or diagnosing a complex industrial control system failure.

11.1 Multimeters

A multimeter is a versatile tool used to measure electrical values such as voltage, current, and resistance. It combines several instruments in one, including a voltmeter, ammeter, and ohmmeter, making it essential for electrical testing.

It is the electrician's primary diagnostic instrument—the stethoscope of the electrical world. With it, you can determine whether circuits are energized, loads are operating correctly, and connections are sound. Modern multimeters can evaluate a wide range of electrical parameters, allowing you to diagnose problems with precision rather than guesswork.

Beyond basic measurements, today's multimeters often include capabilities for testing capacitors, diodes, frequency, duty cycle, temperature, and specialized functions for HVAC and automotive applications. This versatility makes the multimeter the first tool many electricians reach for when troubleshooting.

Types of Multimeters

Analog Multimeters:

Analog multimeters use a needle and scale to display readings. They excel at showing fluctuating values and trends that might be missed with digital sampling. Many can measure basic voltage without batteries, making them reliable backups when digital meters fail. Their multiple overlapping scales require careful interpretation to avoid misreading values. Analog meters typically offer greater resilience in environments with high electromagnetic fields, though their lower input impedance can sometimes affect circuit operation during measurement. While generally less expensive than digital alternatives, they demand more skill to read accurately, especially when interpolating between scale markings.

Digital Multimeters (DMMs)

Digital multimeters display numerical readings on a screen and eliminate the errors that are often associated with analog scales. Most feature automatic ranging that selects the appropriate measurement scale, simplifying operation. Their high input impedance (typically 10MΩ) prevents loading down circuits during testing, yielding more accurate voltage readings. Advanced DMMs include functions like min/max recording to capture intermittent problems, data hold to freeze readings for recording, and relative measurement for comparison testing. The audible continuity function allows testing connections without watching the display. Available in various safety ratings from CAT II for household electronics to CAT IV for utility-level work, DMMs range from $20 for basic models to $500+ for precision industrial instruments with logging capabilities and specialized measurement functions.

11.2 Clamp Meters

A clamp meter, also known as a clamp ammeter, is a tool used to measure current without having to disconnect the circuit. It works by clamping around a conductor and measures the magnetic field generated by the current flowing through it.

This non-invasive measurement method solves one of the biggest challenges in electrical testing: measuring current without breaking the circuit. The clamp meter's jaws contain a current transformer that detects the magnetic field created by flowing electricity. This allows you to measure current in live circuits safely and quickly —simply open the jaws, place them around a single conductor, and read the display.

Types of Clamp Meters

AC Clamp Meters

Designed to measure alternating current (AC).

These basic clamp meters work on standard alternating current circuits like those found in residential and commercial buildings. They typically measure current from 0.1A up to 1000A or more, depending on the model. AC clamp meters use inductive technology, making them more affordable than models that measure both AC and DC.

AC/DC Clamp Meters

Can measure both AC and DC, making them more versatile.

These meters use Hall Effect sensors to detect both alternating and direct current. This versatility makes them essential for automotive work, solar installations, battery systems, and industrial applications where DC measurement is necessary. AC/DC models typically cost more than AC-only clamp meters but offer greater functionality.

True RMS Clamp Meters

Provide more accurate measurements for non-sinusoidal or distorted waveforms.

True RMS (Root Mean Square) meters calculate the effective heating value of current, providing accurate readings even when measuring non-linear loads like variable frequency drives, electronic ballasts, and computer power supplies. Standard "average responding" meters can be off by 40% or more when measuring these distorted waveforms. For industrial work or troubleshooting modern electronic equipment, True RMS capability is essential for accurate measurements.

11.3 Megohmmeters

A megohmmeter, also known as an insulation resistance tester, is used to measure the insulation resistance of electrical components and cables. This tool identifies faulty insulation that may cause short circuits or electrical leakage.

Unlike standard multimeters that measure resistance in ohms, megohmmeters measure extremely high resistance values in megohms (millions of ohms) or gigohms (billions of ohms). They accomplish this by applying a high DC test voltage—typically 250V, 500V, 1000V, or higher—to the insulation being tested. This high voltage stresses the insulation, revealing weaknesses that wouldn't appear under normal circuit voltages.

Modern megohmmeters feature digital displays, multiple test voltages, and safety features that prevent accidental exposure to high voltage. They range from handheld portable units for field use to sophisticated bench models for laboratory testing.

Why Use a Megohmmeter?

Safety
To test the integrity of insulation and prevent electrical shocks.

Insulation breakdown creates shock hazards when conductive parts become energized. By detecting degradation before complete failure occurs, megohmmeters help prevent potentially fatal accidents. This is especially important for equipment operating at higher voltages, where insulation failure can have catastrophic consequences.

Regular testing also identifies moisture ingress, contamination, or physical damage that compromises insulation. These conditions might not cause immediate failure but create dangerous situations where equipment could become energized under certain conditions.

Preventative Maintenance
It ensures the safety and longevity of electrical systems by identifying insulation issues before they cause damage or failure.

Trending insulation readings over time reveals gradual deterioration long before catastrophic failure. Motors, transformers, and cables often show decreasing insulation resistance as they age or operate in harsh environments. This deterioration may not affect normal operation initially, but eventually leads to shorts, ground faults, or complete failure.

By detecting these changes early, maintenance can be scheduled during planned downtime rather than responding to emergency breakdowns. This proactive approach extends equipment life and prevents collateral damage when failures occur.

Compliance

Regular insulation testing is required by various industry standards and regulations.

Many industry standards mandate periodic insulation testing:

- NFPA 70B recommends annual testing for critical motors and equipment.
- ANSI/NETA maintenance standards specify insulation testing procedures and acceptance values.
- Manufacturer warranties often require documented insulation testing to maintain coverage.

These standards exist because insulation testing has proven effective at preventing failures and electrical accidents. Documented testing provides liability protection and demonstrates due diligence in maintaining safe electrical systems.

11.4 Troubleshooting Electrical Faults

Diagnosing electrical problems is more science than art, though it might not seem that way when you're staring at a dead circuit with no obvious issues. Successful troubleshooting comes down to a methodical approach that narrows possibilities until the fault reveals itself.

Experienced electricians know that random part-swapping wastes time and money. Instead, they rely on systematic testing and logical deduction to pinpoint problems. Whether you're facing a tripped breaker in a residence or a complex control system failure in a factory, the same principles apply: gather information, form hypotheses, test systematically, and verify your solution.

In this section, we'll explore proven troubleshooting methods that help you quickly identify electrical faults without guesswork. These approaches work for everything from simple lighting circuits to sophisticated industrial systems.

11.5 Understanding Electrical Faults

Electrical faults can occur in any circuit and are typically caused by issues such as short circuits, open circuits, ground faults, or overloads. Troubleshooting these faults involves systematically isolating the problem to ensure a quick and accurate repair.

When electrical systems fail, they typically do so in predictable ways. By recognizing the characteristics of different fault types, you can narrow down possible causes and test more efficiently. The goal is to identify the exact nature and location of the fault without wasting time on trial-and-error approaches.

Common Electrical Faults

Short Circuits

When a live wire touches a neutral or ground wire, creating an unintended path for current to flow.

Short circuits create low-resistance paths that allow excessive current to flow. This typically trips breakers or blows fuses immediately due to the high current levels. Short circuits generate heat, sometimes creating burn marks or melted insulation at the fault location.

Common causes include damaged insulation, moisture ingress, metal objects across terminals, or improper wiring. These faults can occur in cables, junction boxes, switches, or within equipment itself. The direct connection between conductors creates a path of minimal resistance that can draw maximum available current.

Open Circuits

When a break or disconnection occurs in the wiring, preventing current from flowing.

Open circuits completely stop current flow and result in devices losing all their power entirely. Unlike short circuits that trip breakers, open circuits simply prevent equipment from operating with no visible signs of trouble. This makes them particularly challenging to locate.

Common causes include broken wires, loose terminal connections, failed components, or activated disconnect switches. Open circuits often develop at connection points due to vibration, thermal cycling, or corrosion. Testing for voltage presence at various points helps isolate where the circuit path is interrupted.

Ground Faults

When electricity flows through an unintended path to the ground, often caused by faulty insulation or connections.

Ground faults occur when current finds an alternative path to ground rather than returning through the intended neutral conductor. These faults may trip GFCIs or ground fault protection in breakers, or may manifest as "mysterious" current loss in systems without ground fault protection.

Common causes include moisture on electrical connections, damaged insulation contacting grounded surfaces, or miswired equipment. Ground faults can be dangerous because they may energize conductive surfaces without drawing enough current to trip standard overcurrent protection, creating shock hazards.

11.6 Troubleshooting Process

Identify the Problem

Use your testing equipment (multimeter, clamp meter, megohmmeter) to determine the nature of the fault.

Gather information about the symptoms. What exactly isn't working? When did the problem start? Were there any unusual events (power outage, lightning, recent work) before the failure? This initial assessment helps form theories about potential causes.

Use visual inspection first —look for obvious signs of damage like burn marks, melted insulation, or loose connections. Then apply your test equipment strategically. Check for voltage at various points to see where it's present and where it's missing. Measure current draw to identify overloads or unusual consumption patterns. Test for resistance abnormalities that might indicate damaged components.

The symptom pattern often indicates specific fault types. No power anywhere typically suggests an open circuit or tripped protection. Intermittent operation might indicate loose connections or thermal issues. Equipment that trips breakers immediately suggests a short circuit.

Isolate the Fault

Work through the circuit systematically, checking each component to locate the source of the problem.

Use the divide-and-conquer approach. Test the midpoint of the affected circuit. If voltage is present there, the problem is likely downstream; if not, look upstream. This binary search method quickly narrows down the location of faults in complex systems.

For branch circuits, test at the panel first, then at the first outlet or junction box, working outward toward the end of the circuit. For motor circuits, check incoming power, then controls, then the motor itself. Always move methodically rather than randomly testing components.

Document your findings as you go. Note voltage readings, resistance measurements, and current draws at different points. This creates a map of the circuit's condition that often reveals patterns pointing to the fault location.

Test Continuity

Check for open circuits or damaged connections using your multimeter's continuity setting.

After de-energizing the circuit, use continuity testing to verify the integrity of conductors and connections. This function instantly confirms whether a complete path exists through wires, switches, and components.

Test each conductor run separately to identify breaks. For switches and controls, check continuity in both positions or states. For receptacles and fixtures, verify continuity between termination points. The absence of continuity where a path should exist reveals open circuits or failed components.

Continuity testing helps identify intermittent problems by wiggling wires or connectors while testing. If the continuity beep changes or flickers during movement, you've found a loose connection that might cause intermittent failures under load or vibration.

Verify the Fix

After addressing the fault, re-test the system to ensure the issue is resolved.

Never assume that the first repair attempt fixed the problem. Re-test thoroughly to verify the repair addressed both the symptom and the underlying cause. Check voltage levels, current draw, and operation under normal load conditions.

For intermittent issues, monitor the system over time or under varying conditions. Some faults only appear when systems warm up, when certain loads operate simultaneously, or under specific environmental conditions.

Document the repair for future reference. Note what was found, what was replaced, and any measurements taken before and after the repair. This history becomes valuable if similar issues arise later or if you're troubleshooting patterns of failure.

11.6 Quiz: Testing and Troubleshooting Equipment

Put your knowledge to the test with questions that reflect real-world electrical testing and troubleshooting scenarios. Understanding these concepts and proper testing techniques can mean the difference between quickly resolving problems and wasting hours on ineffective approaches.

Multiple-Choice Questions

1. **When measuring voltage with a multimeter, how should the probes be connected to the circuit?**

   a) In series with the circuit
   b) In parallel across the measurement points
   c) To ground only
   d) With the circuit de-energized

2. **What is the appropriate test voltage when using a megohmmeter on a 230V motor?**

   a) 100V
   b) 250V
   c) 500V
   d) 1000V

3. **When using a clamp meter to measure current in a power cord containing both hot and neutral conductors, what reading would you expect?**

   a) The sum of the current in both conductors
   b) The difference between the currents
   c) Zero or near zero
   d) The average of both currents

4. **Which setting on a multimeter should never be used on an energized circuit?**

   a) AC voltage
   b) DC voltage
   c) Resistance
   d) Frequency

5. **What safety rating should a multimeter have for general electrical work in commercial buildings with 480V systems?**

   a) CAT I 1000V
   b) CAT II 600V
   c) CAT III 600V
   d) CAT IV 300V

True/False Questions

6. **True or False:** When measuring DC voltage, the polarity of the multimeter probes doesn't matter as long as you're willing to interpret negative readings.

7. **True or False:** A clamp meter can accurately measure current in a wire that is bundled with several other current-carrying conductors.

8. **True or False:** A megohmmeter can be used to test insulation resistance while equipment is running under low load.

9. **True or False:** When measuring voltage with a multimeter, you should start with the lowest voltage range and work your way up.

10. **True or False:** A reading of zero ohms on a multimeter always indicates a short circuit.

Scenario-Based Questions

11. **Scenario: You are troubleshooting a circuit where a 20A breaker trips immediately when reset. Your visual inspection shows no obvious damage.**

    What testing approach would you take first?

12. **Scenario: A 480V, three-phase motor won't start. The motor contactor pulls in when the start button is pressed, but the motor just hums without rotating. The breaker doesn't trip.**

    What tests would you perform to diagnose this problem?

13. **Scenario: You need to verify whether a particular conduit run is energized, but you don't have access to the ends of the conductors inside.**

    What alternative testing method might help determine if voltage is present?

14. **Scenario: You're testing a residential circuit where half the outlets in a room have power and half don't.**

    The breaker is on and shows proper voltage at the panel. How would you approach troubleshooting this issue?

15. **Scenario: After an annual megohmmeter test, you find that a motor's insulation resistance has dropped from 100 megohms last year to 20 megohms this year.**

    The motor appears to be running normally. What action, if any, should you take?

Answer Key and Explanations

1. **Answer: b) In parallel across the measurement points**

   Voltage measurements must be taken in parallel with the circuit or component. Connecting a voltmeter in series would effectively open the circuit, preventing normal current flow and yielding inaccurate readings.

2. **Answer: c) 500V**

   For a 230V motor, the standard test voltage would be 500V. The general rule is to use 500V for equipment rated 250V-500V, 1000V for equipment rated 501V-1000V, and 250V for equipment below 250V.

3. **Answer: c) Zero or near zero**

   When clamping around both hot and neutral in the same cable, the current flowing in opposite directions cancels out magnetically, resulting in a zero or near-zero reading. To measure current, the clamp must encircle only one conductor.

4. **Answer: c) Resistance**

   Resistance measurements should never be performed on energized circuits. Applying the resistance function to a live circuit can damage the multimeter and create safety hazards. Always de-energize circuits before resistance testing.

5. **Answer: c) CAT III 600V**

   Commercial buildings with 480V systems require at minimum a CAT III 600V rating, which is designed for distribution-level circuits. CAT II is insufficient for commercial distribution systems, and CAT IV would be necessary only for utility connections.

6. **Answer: True**

When measuring DC voltage, reversed probes will simply give a negative reading. While it's better practice to connect the red probe to the positive point and black to negative, the meter will still provide the correct voltage magnitude, just with a negative sign if reversed.

7. **Answer: False**

A clamp meter measures the magnetic field around current-carrying conductors. When multiple conductors are bundled together, their magnetic fields interact, resulting in inaccurate readings. For accurate current measurement, the clamp must encircle only one conductor.

8. **Answer: False**

Megohmmeters apply high test voltages that can damage equipment and create safety hazards if used on energized systems. Equipment must be completely de-energized, disconnected, and properly discharged before insulation resistance testing.

9. **Answer: False**

When measuring unknown voltages, you should start with the highest voltage range and work down to avoid overloading the meter. Starting too low could damage the meter if the voltage is higher than expected. Many modern autoranging meters handle this automatically.

10. **Answer: False**

A zero-ohm reading could indicate a short circuit, but it could also indicate a properly functioning low-resistance component or connection. Interpretation depends on what's being tested—zero ohms across a switch in the closed position is normal, while zero ohms between unrelated circuits indicates a problem.

11. **Answer:** For a breaker that trips immediately, I would:

* Disconnect loads from the circuit at the first junction or device box
* Test for shorts between hot, neutral, and ground using the resistance function on my multimeter (with power off)
* If no short is found at that point, reconnect and move to the next junction
* Continue this divide-and-conquer approach until I locate the section containing the short
* Look for damaged insulation, pinched wires, or metal contact with conductors

An immediate trip usually indicates a bolted fault (direct short), so I'd focus on finding where conductors might be contacting each other or a grounded surface.

12. **Answer:** For a humming but non-rotating motor, I would:

* Measure voltage at the motor terminals to verify all three phases are present
* Use a clamp meter to check current draw on each phase—an open winding would show zero current on one phase
* Disconnect power and test resistance between each pair of motor leads to check for balanced windings
* Check for mechanical binding by attempting to rotate the shaft by hand (after locking out power)
* Verify motor starter contacts are all closing properly

This pattern often indicates a lost phase (single-phasing) where the motor receives power on only two of three phases, causing the humming without rotation.

13. **Answer:** When conductors aren't accessible, I could:

* Use a non-contact voltage tester along the conduit run—these can often detect voltage through metal conduit
* Use a proximity voltage detector or "tick tracer" near the conduit
* Use an inductive voltage sensor that detects magnetic fields from current-carrying conductors
* Check for warming of the conduit, which might indicate current flow
* As a last resort, use specialized conduit tone generators and tracers

Safety note: These methods only indicate possible presence of voltage—they don't confirm absence of voltage, which requires direct contact testing.

14. **Answer:** For a circuit with partial power loss, I would:

* Determine the pattern of working vs. non-working outlets to establish if they're in sequence
* Check the last working outlet and first non-working outlet in the sequence
* Test for voltage at both hot and neutral terminals at the first non-working outlet
* If voltage is present at the hot terminal but the outlet doesn't work, test for continuity between neutral and ground (which would indicate an open neutral)
* De-energize the circuit and check connections in the junction box between the last working and first non-working outlets

This symptom pattern typically indicates an open neutral connection or a failed connection at a "middle" outlet where the circuit continues to downstream outlets.

15. **Answer:** For declining insulation resistance, I would:

* Document the new reading and compare it to the manufacturer's minimum

specifications

* Schedule more frequent monitoring to establish the rate of degradation
* Check for environmental factors that might have changed (increased humidity, dust, chemical exposure)
* Consider cleaning and drying the motor if moisture ingress is suspected
* Plan for eventual motor replacement or rewind if the trend continues, but 20 megohms is still acceptable for most motors

While 20 megohms is above the typical minimum (1 megohm per kV rating plus 1 megohm), the 80% drop indicates developing issues that warrant monitoring and preventive action before failure occurs.**

Assessment

Count your correct answers:

- 13-15 correct: Excellent understanding of testing and troubleshooting principles
- 10-12 correct: Good knowledge with some areas to strengthen
- 7-9 correct: Basic understanding with significant knowledge gaps
- Below 7: Review needed before performing electrical testing independently

Remember that effective troubleshooting combines technical knowledge with systematic testing methods. The best electricians aren't necessarily those who know the most facts, but those who can apply testing principles logically to isolate and identify problems efficiently.

12

Wiring Techniques and Installations

The lights stay on in a hospital operating room because someone ran those circuits right the first time. A family sleeps peacefully because their service was installed by someone who understood the difference between "minimum requirements" and "what I'd want in my own home." Twenty years after installation, a factory machine still runs reliably because someone took the time to make every termination as if their reputation depended on it.

It did. And so does yours.

Wiring techniques are about creating systems that stand the test of time, maintenance, and the unexpected. The quality of your workmanship lives on long after you've moved to the next job as a testament to your skill and professional integrity.

I've seen so many installations where the wires were connected and the lights came on—initially. But when summer heat expanded connections that weren't properly torqued, when vibration loosened terminals that weren't secured correctly, when water found its way into raceways that weren't properly sealed—that's when the difference between adequate and excellent becomes painfully clear.

This chapter focuses on the techniques that turn code-minimum requirements into installations you'd be proud to show other professionals decades later. We'll explore the methods that experienced electricians rely on to ensure their work remains safe, functional, and serviceable throughout its lifespan.

Because true craftsmanship lives in the details no one can see, until those details become the reason everything works when it matters most.

12.1 Residential Wiring

Purpose

Primarily for homes and small buildings, designed to meet lower electrical demand.

Residential wiring creates the electrical infrastructure that powers our everyday lives at home. These systems typically serve 100-200 amp services distributed through relatively straightforward branch circuits. The design prioritizes safety, reliability, and serviceability while keeping costs reasonable for homeowners.

Residential systems must accommodate diverse loads—from sensitive electronics to major appliances—while remaining simple enough for homeowners to understand basic operation and maintenance. Unlike commercial installations where trained maintenance staff might be available, residential systems need to be intuitive enough for homeowners to operate safely.

The typical residential electrical system contains lighting circuits, general-purpose receptacle circuits, dedicated appliance circuits, and a few specialized circuits like those for HVAC equipment. While homes increasingly include more sophisticated electrical components like renewable energy systems or smart home technology, the fundamental infrastructure remains simpler than commercial installations.

12.2 Types of Wiring

Typically uses non-metallic sheathed cable (NM or Romex) for smaller-scale wiring installations.

Non-metallic sheathed cable dominates residential wiring because it offers several advantages in this environment:

- quick installation without specialized tools
- flexibility for routing through framing members
- lower material costs
- easier modification during renovation

This cabling typically contains two or three insulated conductors plus a bare equipment grounding conductor, all wrapped in a protective outer jacket. Common types include 14/2 for 15-amp circuits, 12/2 for 20-amp circuits, and various other configurations for specialized applications.

Where greater protection is needed, such as outdoor installations or underground runs, specialized cables like underground feeder (UF) or service entrance (SE) cable might be used. In some jurisdictions or specific applications like Chicago, metal conduit may be required even in residential settings.

Wiring Techniques

Simpler wiring methods, often run along studs or in attics and basements, with a focus on ease of installation and cost-effectiveness.

Residential wiring techniques emphasize practicality and efficiency. NM cable typically follows framing members, secured with staples at prescribed intervals. Proper installation includes:

- drilling holes in studs and joists at the appropriate height
- maintaining minimum distances from the edges of framing
- using protective plates where required to prevent nail or screw penetration
- allowing proper slack at boxes and fixtures

Unlike commercial installations, where accessibility for future modification is paramount, residential wiring often becomes enclosed within finished walls. This makes initial planning critical, as future access typically requires opening wall surfaces.

Junction boxes must remain accessible after construction, though they're often placed in less conspicuous locations like closets or basements. Careful labeling of circuits in the panel helps future troubleshooting when problems arise.

Load Considerations

Electrical systems in residential properties have lower amperage demands and simpler configurations.

A typical modern home might have a 200-amp service supporting various loads:

- General lighting and receptacle circuits at 15 or 20 amps
- Dedicated 20-amp circuits for kitchens and laundry
- 30-50 amp circuits for electric ranges or dryers
- 20-60 amp circuits for HVAC equipment

These systems typically operate at single-phase 120/240 volts. Load calculations follow NEC Article 220, with standard allowances for general lighting, small appliance circuits, and factored calculations for larger loads.

Power quality concerns are generally less critical in residential settings, though surge protection has become increasingly important as homes contain more sensitive electronics. Most residential systems don't require power factor correction or harmonics mitigation that might be necessary in commercial environments.

The relative simplicity of residential systems doesn't diminish their importance—these installations must still provide decades of safe, reliable service while accommodating changing usage patterns as homes evolve.

12.3 Commercial Wiring

Purpose

For businesses, industrial spaces, and larger buildings, designed to handle higher electrical loads.

Commercial wiring systems power environments where electrical failures can have significant economic impact. From small retail stores to massive industrial complexes, these systems must deliver reliable power under demanding conditions while allowing for future expansion, reconfiguration, and maintenance access.

Commercial electrical systems must accommodate diverse equipment needs, support life safety systems, and continue functioning through decades of changing tenants or production requirements. These installations prioritize durability, serviceability, and flexibility over lowest initial cost.

The design often incorporates greater redundancy, more robust components, and systematic organization to support maintenance and troubleshooting by facility personnel. Unlike residential systems, where aesthetics may be a primary concern, commercial installations focus on accessibility, longevity, and performance under continuous operation.

Types of Wiring

Often uses conduit-based systems such as EMT (Electrical Metallic Tubing), IMC (Intermediate Metal Conduit), or PVC (Polyvinyl Chloride).

- Raceway-based wiring methods dominate commercial installations for several reasons:
- enhanced physical protection for conductors
- greater capacity for future modification
- superior protection in harsh or hazardous environments
- improved support for larger conductor sizes
- better heat dissipation for heavily loaded circuits

EMT (Electrical Metallic Tubing) is the most common raceway in commercial buildings, offering good protection at reasonable cost. Where greater protection is needed, Intermediate Metal Conduit (IMC) or Rigid Metal Conduit (RMC) may be used. Non-metallic raceways like PVC find applications in corrosive environments or underground installations.

Individual THHN/THWN conductors pulled through these raceways offer greater flexibility than pre-assembled cables. This approach allows damaged conductors to be replaced without removing entire cable assemblies and facilitates adding conductors when needs change.

Cable trays provide organized support for multiple cables in data centers, industrial facilities, and other applications where many conductors must run parallel paths. MC (Metal-Clad) cable offers a middle ground between conduit systems and NM cable, finding applications in commercial settings where permitted by code and specifications.

Wiring Techniques

More complex, often requiring professional skills to run wires through conduit, use of cable trays, and multiple circuit configurations.

Commercial wiring requires planning and techniques beyond residential applications:

- bending conduit to precise measurements and angles
- calculating and respecting maximum conduit fill
- maintaining appropriate support intervals for various raceway types
- planning for expansion and pull points in long conduit runs

Installations often incorporate diverse distribution methods, from traditional conduit to busways, cable trays, and cellular floor systems. Each requires specific installation techniques and knowledge of code requirements.

Conductor management becomes critical in panels and gutters, with phase identification, clear labeling, and organized routing. Panel schedules must be accurate and detailed to support future maintenance and modifications.

Commercial installations require careful planning for equipment mounting, working clearances, and access requirements. Coordination with other building systems—structural, mechanical, plumbing—becomes essential to prevent conflicts during construction.

Load Considerations

Designed to meet higher amperage requirements, with more circuits and outlets, and sometimes multiple phases of power.

Commercial electrical systems typically feature:

- three-phase power distribution (commonly 208Y/120V or 480Y/277V)
- higher capacity services (400A to several thousand amps)
- more sophisticated distribution equipment (switchgear, panel boards, transformers)
- higher circuit densities to support concentrated equipment loads

Load calculations follow NEC Article 220 but typically incorporate more detailed analysis of specific equipment requirements, diversity factors, and demand considerations. Power quality concerns receive greater attention, with harmonic mitigation, power factor correction, and transient voltage protection becoming standard considerations.

Emergency and standby power systems play a more prominent role in commercial installations, with separate distribution paths for life safety, legally required standby, and optional standby loads. These systems require specialized knowledge of transfer equipment, separation requirements, and testing protocols.

The complexity of commercial systems creates both challenges and opportunities for electricians to develop advanced skills in distribution design, equipment selection, and system integration.

12.4 Conduit Bending and Wire Pulling Techniques

If you talk to any seasoned electrician about their most valuable skills, conduit bending often tops the list. There's an unmistakable pride in walking through a mechanical room where perfectly bent conduit runs parallel to building lines, offsets match perfectly, and every bend shows precision rather than guesswork.

The art of bending conduit and pulling wire separates professionals from amateurs in our trade. Anyone can connect point A to point B with enough elbows and couplings, but creating efficient, maintainable installations requires understanding both the mechanical principles behind conduit bending and the practical realities of wire pulling.

These skills develop through practice, mentorship, and experience. While some aspects can be learned from books, the feel of a properly executed bend and the techniques for pulling wire through complex runs come from hands-on repetition.

12.5 Conduit Bending Basics

Conduit bending is a critical skill for creating neat, safe, and efficient electrical installations. It ensures the conduit conforms to the design of the system while protecting the wires inside.

Proper conduit installations require not just connecting endpoints but creating systems that follow building lines, avoid obstacles, maintain appropriate support intervals, and allow for efficient wire pulling. Each bend introduces friction and potentially complicates future maintenance, so thoughtful planning and precise execution are essential.

The fundamental bending techniques—90-degree bends, offsets, saddles, and kicks—form the building blocks of conduit installations. Mastering these basic bends and understanding how to combine them allows electricians to navigate complex pathways while maintaining professional appearance and functionality.

Types of Conduit

EMT (Electrical Metallic Tubing)

They're lightweight, easy to bend, and widely used in commercial and residential applications.

EMT is the most common conduit type in commercial construction due to its balance of protection, cost, and ease of installation. This thin-walled steel tubing bends readily with hand or mechanical benders while providing good protection from physical damage.

It is available in sizes from 1/2-inch to 4-inch diameter. EMT works well for most indoor applications where severe physical damage isn't expected. Its smooth interior surface facilitates wire pulling, while its metallic construction provides an equipment grounding path when properly installed with appropriate fittings.

EMT requires protection from corrosive environments and isn't suitable for direct burial. It connects using set-screw, compression, or indenter-type fittings depending on the environment and local code requirements.

IMC (Intermediate Metal Conduit)

Slightly heavier than EMT but more robust, offering better protection for wiring in industrial settings.

IMC provides a middle ground between lightweight EMT and heavy-duty rigid metal conduit. Its thicker walls offer greater protection from physical damage, and make it appropriate for more demanding installations.

This conduit type requires more force to bend, especially in larger sizes, often necessitating hydraulic or mechanical benders rather than hand benders. The thicker walls also mean IMC maintains its circular shape better during bending, reducing the risk of flattening that can complicate wire pulling.

IMC finds applications in industrial settings, areas subject to moderate physical damage, and environments where the added strength provides value. Its threaded connections offer superior pull-out strength compared to typical EMT fittings.

PVC (Polyvinyl Chloride)

This one is flexible and resistant to corrosion, often used for underground or wet-location installations.

PVC conduit offers excellent corrosion resistance, making it ideal for wet locations, direct burial, or chemically harsh environments. Unlike metallic conduits that are cold-bent, PVC requires heating to become pliable enough for bending.

Heating methods include specialized "hot boxes" or heating blankets that warm the conduit evenly to bending temperature. The bending process requires proper heating (not too hot or too cool) and supporting the conduit during cooling to prevent flattening or spring-back.

PVC conduit must include a separate equipment grounding conductor since the conduit itself doesn't provide a reliable ground path. Schedule 40 PVC serves most applications, while Schedule 80 provides greater impact resistance for areas subject to physical damage.

Each conduit type requires specific techniques, tools, and fittings. The electrician's skill lies not just in executing bends but in selecting the appropriate conduit type and installation method for each application.

12.6 Bending Techniques

Determine the Bend Angle

Measure the required angle for your installation using the appropriate formula or bender scale.

Accurate measurement is the foundation of successful conduit bending. Before making any bend, you must determine the exact dimensions needed. Common calculations include:

For 90-degree bends, calculate the distance to the back of the bend by subtracting the deduct value for your bender. This compensates for the gain that occurs during bending. Each bender has specific deduct values based on conduit size and bender design.

For offsets, determine the depth of offset needed and calculate the distance between bends. The formula is typically: Distance between bends = Offset depth ÷ sine of the bend angle. Using smaller angles (like 10, 22, or 30 degrees) for offsets creates more gradual transitions and makes wire pulling easier than sharper 45 or 60-degree bends.

For saddles, calculate both the height needed to clear obstacles and the distance between bends. A typical three-point saddle uses a center bend of twice the angle of the two outer bends, maintaining the conduit's original direction while navigating over obstacles.

Place Conduit in Bender

Position the conduit securely in the bender and align it with the correct markings.

Proper alignment ensures accurate bends. Insert the conduit into the bender's groove, ensuring it seats fully against the roller or hook. Align your measured mark with the appropriate arrow on the bender —different arrows correspond to different types of bends.

For hand benders, position your foot securely on the foot pedal and maintain firm control of the handle. Your non-bending hand should hold the conduit securely to prevent slipping during the bend.

For mechanical or hydraulic benders, secure the conduit according to the manufacturer's instructions, typically using shoes specific to the conduit size. Verify alignment before applying pressure, as corrections become difficult once bending begins.

Apply Even Pressure

Apply slow and steady pressure on the bender to create the desired bend, ensuring no kinks or distortions.

The bending motion should be smooth and continuous to prevent flattening or kinking. With hand benders, maintain downward pressure on the handle while controlling the bend rate. Too much force applied too quickly often results in kinked conduit.

For mechanical benders, operate the controls at a moderate pace, observing the conduit as it bends. Stop periodically to check progress on precision bends rather than completing the bend in one motion.

Watch for signs of flattening or crimping during bending, especially with EMT. If you notice the conduit beginning to deform, stop immediately, as continuing will likely result in unusable conduit. Larger sizes typically require more gradual pressure to prevent flattening.

Check the Bend

After bending, verify the angle and alignment to ensure it fits the installation plan.

Accurate measurement after bending confirms success before proceeding. For 90-degree bends, use a square to verify the angle. For offsets and saddles, measure the rise and compare it to your intended dimensions.

Place the bent conduit against the installation path to verify fit. Minor adjustments can sometimes be made by hand for small-diameter conduit, but significant errors typically require rebending with new material.

Maintain a one-plane bend whenever possible, ensuring the entire bend occurs in the same plane. Multi-plane bends create additional complexity for wire pulling and often indicate poor planning or execution.

For complex runs with multiple bends, verify that the total does not exceed 360 degrees between pull points, as required by the NEC. Excessive bends dramatically increase pulling difficulty and may necessitate additional pull boxes.

Mastering these bending techniques requires practice and experience. Many electricians keep bending calculation apps or cards handy to reference common formulas and deduct values specific to their bending equipment.

12.7 Wire Pulling Considerations

The most perfectly bent conduit system becomes useless if wires can't be pulled through it efficiently. Successful wire pulling requires planning, proper equipment, and techniques that minimize conductor damage while allowing for future changes.

Conduit Size

The conduit must be large enough to accommodate the wires, allowing for easy pulling and avoiding excess tension.

- Conduit fill calculations determine the maximum number of conductors permitted in a raceway, typically limited to 40% of the conduit's interior cross-sectional area for three or more conductors.
- Exceeding recommended fill percentages dramatically increases pulling tension and the risk of conductor damage, potentially violating code requirements and creating future

failure points.

- Future expansion needs should influence initial sizing decisions. A slightly larger conduit often pays dividends when systems require modification or additional circuits.
- Minimum bend radius requirements increase with conductor size and quantity. Larger conduit allows gentler bends that reduce pulling stress and potential insulation damage.
- High conductor count pulls may require larger conduit than calculations suggest to account for the practical difficulties of pulling many wires simultaneously.

Wire Lubricants

Lubricants can be applied to reduce friction and make wire pulling easier and safer.

- Different lubricant types suit specific applications—some formulated for PVC conduit, others for EMT or large commercial pulls.
- Apply lubricant directly to conductors as they enter the conduit, ensuring even coverage without excessive amounts that could contaminate termination points.
- For longer pulls, consider intermediate lubricant application points using conduit bodies or pull boxes to maintain lubrication throughout the run.
- Temperature affects lubricant performance—cold weather installations may require special low-temperature formulations to maintain effectiveness.
- Water-based lubricants offer easier cleanup but may dry out over time, while silicone or wax-based products provide longer-lasting reduction in friction.
- Calculate expected pulling tension for complex or critical installations to determine if additional lubricant or larger conduit is needed.

Fish Tapes and Pulling Grips

Use fish tapes (a flexible tool used to pull wires through conduit) or pulling grips for efficient wire movement.

- Match fish tape material to the installation—steel tapes work well for EMT but can damage PVC conduit, while fiberglass or nylon tapes reduce risk of damage or electrical issues.
- Secure pull strings during conduit installation to eliminate the need for fish tapes in complex runs—a simple preventative step that saves significant time.
- For heavier pulls, use pulling grips (wire mesh sleeves) that distribute tension across the conductor bundle rather than stressing a single point connection.
- Mechanical wire pullers provide controlled force for larger installations, reducing strain on workers while monitoring pulling tension to prevent conductor damage.
- Stage longer pulls with personnel at feeding points, midpoints, and pulling ends to

coordinate movement and prevent binding or kinking.
- Tape wire bundles in a streamlined fashion, staggering the connections to create a tapered nose that navigates bends more easily than blunt-ended bundles.

Proper pulling techniques preserve conductor insulation integrity and ensure the installation will provide decades of reliable service. Taking time to plan pulls thoroughly and execute them properly prevents the false economy of rushing this critical phase of installation.

12.8 Step-by-Step Wire Pulling Process

Successful wire pulls don't happen by accident. They result from methodical preparation and execution that prevents damage while ensuring conductors reach their destination intact. The following process represents industry best practices for wire pulling in conduit systems.

Preparation

Ensure that the conduit is free of sharp edges and debris.

- Inspect the entire conduit run before beginning, checking for proper support, secure fittings, and completed connections throughout the path.
- Blow compressed air through the conduit to remove any debris, water, or foreign objects that could impede pulling or damage conductors.
- Verify that all fittings are properly tightened and that no burrs or sharp edges exist at conduit ends or connector points.
- Ream all cut ends of conduit to eliminate sharp edges that could damage wire insulation during pulling.
- Pre-lubricate complex or challenging runs, applying lubricant directly to the inside of the conduit before beginning the pull.
- Arrange conductors near the feed point in a manner that prevents tangling or kinking during the pull, potentially using wire spools or stackers for larger pulls.

Pulling the Wire

Feed the wire into the conduit while ensuring it doesn't snag. Use fish tape or a pulling grip to guide it.

- Push the fish tape or pull string through the conduit first, working from the end that requires the fewest bends to reach the opposite termination.
- Secure conductors to the fish tape or pull string using electrical tape in a streamlined,

tapered formation that will navigate bends smoothly.

- For multiple conductors, stagger the connections to create a gradual taper rather than a blunt end that might catch on bends or fittings.
- Apply lubricant to the conductors as they enter the conduit, ensuring even coverage while avoiding excess that might contaminate termination points.
- Maintain communication between team members at feeding and pulling positions, especially on longer runs where visibility between ends is limited.
- Feed conductors into the conduit gradually, working with the puller to maintain smooth movement without jerking or sudden tension changes.

Avoid Over-tensioning

Maintain a steady pull, but avoid putting excessive force on the wire, which can cause damage or overheating.

- Pull with steady, consistent force rather than jerking motions that can damage insulation or break conductors.
- For longer or more difficult pulls, use mechanical advantages like pulley systems rather than excessive manual force.
- Be aware of maximum allowable pulling tension for the specific conductor types being installed – larger conductors can tolerate more pulling force than smaller ones.
- Watch for signs of binding, such as increasing resistance or stopping of movement, which indicate potential problems requiring investigation.
- When resistance increases, stop pulling and identify the cause rather than applying more force, which could damage conductors or conduit systems.
- Consider using pulling lubricant injection systems for longer runs to maintain lubrication throughout the pull.

Completing the Pull

Once the wire reaches the other end, check for any potential damage, ensuring the wire is intact and the pull was successful.

- Visually inspect exposed portions of the conductors for any signs of insulation damage, stretching, or deformation.
- Leave sufficient conductor length at termination points: typically 6-8 inches in device boxes and 18-36 inches in panels or equipment.
- Label conductors immediately after pulling according to project specifications, using durable identification methods appropriate for the environment.
- Secure conductors temporarily at terminations to prevent them from being pulled back into

the conduit.
- Document any deviations from plans or specifications, including conductor substitutions or routing changes.
- Perform insulation resistance testing on feeder or critical circuit conductors before termination to verify insulation integrity after pulling.

Following this systematic approach to wire pulling helps ensure that electrical systems will function reliably throughout their service life, with minimal risk of insulation failure or connection problems that often result from hasty or improper installation techniques.

12.9 Quiz: Wiring Techniques and Installations

Test your understanding of proper wiring techniques with questions that reflect real-world situations you'll encounter in the field. This quiz covers the critical differences between installation types and the practical skills needed for professional electrical work.

Quiz Overview

- **Multiple-choice questions**: Test your knowledge on residential and commercial wiring differences, conduit bending, and wire pulling techniques.
- **True/False questions**: Assess understanding of NEC compliance and safety.
- **Scenario-based questions**: Apply your knowledge to common wiring installation scenarios to ensure compliance with NEC standards.

Multiple-Choice Questions

1. **What is the maximum fill percentage allowed when installing three or more current-carrying conductors in a conduit?**

   a) 25%
   b) 31%
   c) 40%
   d) 53%

2. **When bending a 90-degree bend in ½" EMT conduit, what must be accounted for in your measurements?**

    a) The gain in the overall conduit length
    b) The deduct value of the bender
    c) The conduit shrinkage during bending
    d) The radius multiplier factor

3. **Which wiring method is NOT typically permitted in commercial kitchen areas?**

    a) EMT conduit
    b) Non-metallic sheathed cable (NM/Romex)
    c) Metal-clad cable (MC)
    d) Rigid metal conduit (RMC)

4. **What is the primary reason for using wire pulling lubricant?**

    a) To increase the current-carrying capacity of the conductors
    b) To reduce friction and prevent insulation damage during pulling
    c) To protect wires from moisture after installation
    d) To improve electrical connectivity at termination points

5. **In a residential application, what is the minimum burial depth required for direct-buried UF cable without concrete protection?**

    a) 6 inches
    b) 12 inches
    c) 18 inches
    d) 24 inches

True/False Questions

6. **True or False:**

    Non-metallic sheathed cable (NM/Romex) can be used in all types of commercial buildings as long as it's not exposed.

7. **True or False:**

    When pulling wire, it's acceptable to use more force if the conductors become stuck, as modern insulation is designed to withstand high pulling tension.

8. **True or False:**

   For EMT conduit installations, the total number of bends between pull points cannot exceed 360 degrees.

9. **True or False:**

   Wire nuts are an acceptable method for splicing conductors in all commercial applications.

10. **True or False:**

    When bending PVC conduit, heating the conduit evenly is critical to prevent flattening and ensure proper bends.

Scenario-Based Questions

11. **Scenario:** You are installing a 120-foot run of 2-inch EMT conduit with three 90-degree bends for a feeder circuit. The run will contain six 3/0 THHN conductors. What considerations should guide your installation approach for successful wire pulling?

12. **Scenario:** You are wiring a residential garage workshop. The homeowner wants to install a 240V, 30A circuit for a welder, along with several 120V receptacles for power tools. Describe the appropriate wiring methods, materials, and NEC requirements for this installation.

13. **Scenario:** During a commercial renovation, you discover that the existing EMT conduit system needs to be extended to reach new equipment locations. The existing conduit is nearly at maximum fill capacity. What options should you consider, and what NEC provisions would guide your approach?

14. **Scenario:** You need to create a 6-inch offset in a run of 3/4" EMT to avoid a structural beam. Describe the proper procedure for measuring, marking, and bending this offset, including the formulas or calculations you would use.

15. **Scenario:** You're installing a new service for a residence. The service entrance conductors need to be routed through the foundation wall below grade. What methods and materials would be appropriate for this installation, and what specific NEC requirements would apply?

Answer Key and Explanations

1. **Answer: c) 40%**

   For three or more conductors, NEC Table 1, Chapter 9 limits conduit fill to 40% of the conduit's internal cross-sectional area. This ensures adequate space for pulling without damaging insulation and allows for heat dissipation during operation.

2. **Answer: b) The deduct value of the bender**

   When making a 90-degree bend, you must subtract the bender's deduct value from your measured distance to account for the gain in length that occurs during bending. Each bender has specific deduct values based on conduit size and bender design.

3. **Answer: b) Non-metallic sheathed cable (NM/Romex)**

   NM cable is not permitted in commercial kitchens due to the requirements for more robust wiring methods in these locations. Commercial kitchens are typically considered damp locations with exposure to cleaners and chemicals. NEC 334.10 restricts NM cable to dry locations and specific building types.

4. **Answer: b) To reduce friction and prevent insulation damage during pulling**

   Wire pulling lubricant reduces friction between conductors and conduit walls, lowering pulling tension and preventing insulation damage. It does not affect current capacity, moisture protection, or electrical connectivity.

5. **Answer: d) 24 inches**

   NEC Table 300.5 requires a minimum burial depth of 24 inches for direct-buried UF cable without concrete protection in residential applications. Lesser depths are permitted only with specific protection methods like concrete encasement.

6. **Answer: False**

   NEC 334.10 restricts NM cable use based on building type and construction, not just exposure. NM cable is generally not permitted in many commercial buildings, particularly those exceeding three floors above grade or classified as other than Type III, IV, or V construction.

7. **Answer: False**

   Excessive pulling force can damage conductor insulation and create future failure points, regardless of insulation type. When conductors become stuck, the proper response is to investigate the cause rather than applying more force.

8. **Answer: True**

   NEC 358.26 limits the total combined degrees of bends between pull points to 360 degrees (four 90-degree bends) for EMT conduit. This restriction applies to all types of raceways to prevent excessive pulling tension during wire installation.

9. **Answer: False**

   While wire nuts are acceptable for many applications, certain commercial installations require more robust connection methods like split-bolt connectors, compression connectors, or terminal blocks, particularly for larger conductors or critical circuits.

10. **Answer: True**

    PVC conduit must be heated evenly to ensure uniform pliability during bending. Uneven heating causes some areas to bend more sharply than others, resulting in flattening or kinking that obstructs wire pulling and compromises the installation.

11. **Answer:** This installation presents several challenges requiring careful planning:

    * Conduit fill calculation must verify that six 3/0 THHN conductors do not exceed the 40% maximum fill for 2-inch EMT
    * With three 90-degree bends totaling 270 degrees, the installation approaches the 360-degree maximum but remains compliant
    * The 120-foot length with multiple bends will create significant pulling tension, requiring:

      » Liberal use of appropriate wire pulling lubricant
      » Possibly staged pulling techniques with access at intermediate points
      » Proper pulling equipment rated for the expected tension
      » A wire pulling grip or basket to distribute tension across all conductors

        › Conductors should be pulled simultaneously rather than individually to prevent damage to the first conductors from repeated pulls
        › Verify all conduit connections are secure and all cut ends are properly reamed before pulling

    —Consider slightly upsizing the conduit beyond minimum requirements to reduce pulling tension

    This installation would benefit from planning pull direction to minimize tension through bends and ensuring adequate manpower for feeding and pulling operations.

12. **Answer:** For a residential garage workshop with 240V and 120V circuits:

* Appropriate wiring methods include NM cable (if the garage is dry) or MC cable. Conduit systems like EMT are also acceptable and provide greater protection.
* For the 240V, 30A welder circuit:

  » Use 10 AWG copper conductors with appropriate insulation (typically THHN in conduit or NM/MC cable)
  » Install a dedicated 30A, 2-pole circuit breaker in the panel
  » Use a NEMA 6-30R receptacle rated for 240V, 30A service
  » Ensure the receptacle height meets NEC requirements (typically 15-48 inches above floor)
  » For 120V receptacles:

    › Use 12 AWG copper conductors for 20A circuits or 14 AWG for 15A circuits

  » Install GFCI protection for all garage receptacles per NEC 210.8(A)(2)
  » Ensure receptacle spacing meets NEC 210.52(G) requirements
  » Consider dedicated 20A circuits for power tools to prevent nuisance tripping

    › All circuits require proper grounding and must comply with NEC Article 250
    › The installation must follow NEC 210.11(C)(3) if the garage is attached to the dwelling
    › If the garage is detached, consider proper feeder sizing and separate grounding requirements per NEC 250.32

This installation provides both the high-power 240V circuit needed for welding and adequate general-purpose receptacles for workshop tools, while maintaining NEC compliance for safety.**

13. **Answer:** When extending an existing conduit system that's near maximum fill:

* First, verify the actual fill percentage of the existing conduit by identifying all conductors and calculating the total area using NEC Chapter 9, Table 5
* Options to consider include:

  » Installing a parallel conduit run for the new circuits, leaving the existing conduit as is
  » Replacing the existing conduit with a larger size that can accommodate both existing and new conductors
  » Installing a junction box at the extension point and upsizing the new conduit section. The NEC provisions that guide this approach include:

    › 300.17 regarding number and size of conductors in raceway

- › Chapter 9, Table 1 for conduit fill percentages (40% maximum for 3+ conductors)
- › 312.8 regarding junction and pull boxes for wire splicing

» 110.14 for requirements on conductor terminations and splices

- › When making connections at junction points, maintain proper support within 3 feet of boxes per NEC 358.30
- › Ensure accessibility of all junction boxes per NEC 314.29
- › Consider derating factors for conductor ampacity if the number of current-carrying conductors exceeds three (NEC 310.15(C))

This approach balances the practical need to extend the system with code compliance for safety and future serviceability.**

14. **Answer:** To create a 6-inch offset in 3/4" EMT:

* **Step 1:** Determine the bend angle to use (typically 30° for a 6-inch offset provides a gradual transition)
* **Step 2:** Calculate the distance between bends using the formula: Distance between bends = Offset depth ÷ sin(angle). For a 6-inch offset using 30° bends: 6 ÷ sin(30°) = 6 ÷ 0.5 = 12 inches
* **Step 3:** Mark the conduit at the first bend point
* **Step 4:** Place the conduit in the bender with the mark aligned with the 30° mark on the bender
* **Step 5:** Bend to 30° using steady pressure
* **Step 6:** Mark the second bend point 12 inches from the first mark
* **Step 7:** Flip the conduit and place the second mark at the 30° mark on the bender, ensuring the first bend is oriented to create the offset in the same plane
* **Step 8**: Bend the second bend to 30°
* **Step 9:** Check that the offset measures 6 inches and both bends are in the same plane
* **Step 10:** Verify that the overall length of the conduit accounts for the "shrink" that occurs when creating an offset (approximately 3/8" for this offset)

This procedure creates a professional offset that maintains proper conduit alignment while navigating around the structural beam.

15. **Answer:** For service entrance conductors through a foundation wall:

Appropriate methods include:

* Rigid Metal Conduit (RMC)
* Intermediate Metal Conduit (IMC)
* Schedule 80 PVC conduit
* Service Entrance Cable (type USE or SE) with proper protection
* The installation must comply with several NEC requirements:
* NEC 230.8 requires raceways entering a building to be sealed to prevent moisture intrusion
* NEC 300.5(A) requires minimum burial depths (typically 24" for direct burial or 18" with mechanical protection)
* NEC 300.5(C) requires underground raceways entering buildings to be arranged to prevent water entry
* NEC 300.5(G) requires a bushing or equivalent at the end of the raceway where cables enter
* Additional considerations include:
* Proper support of the conduit on both sides of the foundation wall
* Allowing for settling and frost heaving without stressing the conduit
* Using expansion fittings if necessary for significant temperature differentials
* Ensuring the foundation penetration is watertight (typically using hydraulic cement or appropriate sealants)
* Providing adequate drainage away from the service entrance

This installation method ensures a code-compliant, weatherproof entrance for service conductors that will maintain its integrity over the life of the installation.**

Assessment

Count your correct answers:

- 13-15 correct: Excellent understanding of wiring techniques and installations
- 10-12 correct: Good knowledge with some areas to strengthen
- 7-9 correct: Basic understanding with significant knowledge gaps
- Below 7: Review needed before performing independent installations

Remember that proper wiring techniques are the foundation of safe, reliable electrical systems. Mastering these skills takes practice, but the investment pays dividends throughout your career in more efficient installations and fewer callbacks.

13

Electrical Calculations and Load Analysis

Electricity is unforgiving. Unlike plumbing, where a small leak might be tolerated, or carpentry, where a slight measurement error can be sanded away, electrical systems demand precision. Too little capacity and systems fail; too much and you've wasted resources and money. This precision begins with calculations.

Behind every successful electrical installation lies a foundation of careful mathematics. From the residential electrician sizing a service panel for a growing family to the industrial engineer designing power distribution for a manufacturing plant, load calculations determine whether systems will operate safely, efficiently, and economically.

In this chapter, we move beyond theory into the practical mathematics that transforms abstract electrical principles into working systems. These aren't merely academic exercises—they're the tools that prevent overloaded circuits, dangerous voltage drops, and catastrophic failures. They're also the calculations that save money by preventing overbuilt systems with unnecessary capacity.

You'll learn how to analyze existing loads, predict future requirements, and design systems with appropriate safety margins. Whether you're troubleshooting an overloaded circuit in a residential kitchen or planning power distribution for an industrial facility, these calculations form the bedrock of sound electrical practice.

The ability to perform these calculations with confidence separates those who merely follow directions from those who truly understand their craft. Master these skills, and you'll never again be at the mercy of guesswork or rule-of-thumb estimates that might prove inadequate when actual loads materialize.

Load analysis isn't just about numbers—it's about understanding how people and systems use electricity in the real world.

13.1 Conductor Sizing and Ampacity Calculations

The wire that delivers electricity to a device is perhaps the most underappreciated component in any electrical system. Too small, and it becomes the weakest link—heating up, wasting energy, and potentially creating a fire hazard. Too large, and you've wasted money on copper that provides no additional benefit. Finding that perfect balance isn't guesswork; it's science.

Every electrical conductor has its limits—a maximum current it can safely carry without overheating. This limit, known as ampacity, is where electrical theory meets practical application. When a conductor heats up under load, it's telling you something: either it's working within design parameters or it's crying out for help. Learning to calculate, understand, and respect these limits is essential to both safety and efficiency.

I've seen the consequences of undersized conductors: melted insulation, charred panels, and even structural fires. Each incident began with the same mistake—someone who guessed instead of calculated, who hoped instead of knew. These calculations aren't just academic exercises; they're practical tools that protect lives and property.

Let's explore how to determine the right conductor for any application—a skill that will serve you whether you're wiring a simple residential circuit or designing a complex industrial system.

13.2 Conductor Sizing Basics

Conductor sizing ensures that the wiring used in an electrical system can safely carry the amount of current required without overheating or causing a fire hazard. Ampacity refers to the maximum current a conductor can carry before it becomes too hot, which can lead to insulation failure and other safety issues.

Proper conductor sizing is fundamental to electrical safety. Undersized conductors create resistance heating that degrades insulation over time, potentially leading to shorts, fires, or equipment damage. They also cause voltage drop that reduces equipment performance and efficiency. Conversely, grossly oversized conductors waste materials and increase installation costs unnecessarily.

The National Electrical Code provides detailed tables and adjustment factors for determining minimum conductor sizes. These requirements represent safety minimums—many installations benefit from sizing above these minimums for future expansion, efficiency, or reduced voltage drop.

Key Factors in Conductor Sizing

Ampacity

Ampacity tables in NEC Article 310 provide the foundation for conductor sizing. Table 310.16 lists the allowable ampacities for insulated conductors rated up to 2000 volts in ambient temperatures not exceeding 30°C (86°F). The values in these tables assume no more than three current-carrying conductors in a raceway or cable.

When conditions deviate from these assumptions, adjustment factors must be applied. For instance, bundling more than three current-carrying conductors reduces the conductors' ability to dissipate heat, requiring derating. Similarly, ambient temperatures above 30°C reduce allowable ampacity, while temperatures below 30°C sometimes permit higher ampacities.

These ampacity ratings must be respected to prevent insulation degradation that can lead to premature failure, short circuits, or electrical fires over time.

Temperature Rating

Different insulation types (e.g., THHN, XHHW) have different temperature ratings, affecting the ampacity.

Insulation temperature ratings directly impact a conductor's ampacity. Common ratings include:

- 60°C (140°F) for types TW, UF
- 75°C (167°F) for types THW, THWN, XHHW
- 90°C (194°F) for types THHN, XHHW-2, THWN-2

While higher temperature ratings theoretically allow higher ampacities, the NEC limits termination temperatures to the rating of the connected equipment. Most standard equipment terminals are rated for 75°C maximum, meaning that 90°C wire must often be sized based on the 75°C column unless all connected equipment is specifically rated for 90°C terminations.

The temperature rating also affects the conductor's suitability for different environments. Higher-rated insulations generally offer better resistance to heat damage, chemical exposure, and aging effects.

Conductor Material

Copper conductors have a higher ampacity than aluminum for the same size wire.

The choice between copper and aluminum significantly impacts conductor sizing. For a given current requirement, aluminum conductors must typically be one or two sizes larger than copper equivalents due to aluminum's lower conductivity (approximately 61% that of copper).

While aluminum offers cost and weight advantages, particularly in larger sizes, it requires special consideration for terminations. All terminations must use connectors listed for aluminum or copper-aluminum combinations, often with antioxidant compounds to prevent connection degradation over time.

Copper is still dominant in branch circuits and smaller feeders due to its higher conductivity, mechanical strength, and termination reliability. Aluminum finds common use in larger feeders and service conductors where its cost advantages offset the increased size requirements.

Circuit Type

The type of circuit (residential, commercial, etc.) will dictate how conductors are sized.

Different circuit types have specific sizing requirements based on their use:

- Branch circuits typically follow straightforward sizing from NEC 210.19, generally requiring conductors sized at 125% of the continuous load plus 100% of the non-continuous load.
- Feeder circuits must account for load diversity factors and potential future expansion, often resulting in larger conductors than the minimum calculated size.
- Motor circuits have specialized sizing requirements detailed in NEC Article 430, accounting for starting currents, duty cycles, and overload protection strategies.
- Service conductors must be sized according to NEC Article 230, considering the calculated load and any service factors required by the utility provider.

These varied requirements ensure that conductors are properly sized for their specific application, providing safe and reliable operation while meeting code requirements.

13.3 Calculating Ampacity

Properly determining conductor ampacity requires understanding both the basic electrical calculations and the application of NEC tables and adjustment factors. This process ensures conductors can safely carry required current while maintaining proper temperature limits.

Ampacity is determined by using tables from the **National Electrical Code (NEC)**, which specify ampacity for different conductor sizes and materials.

The NEC provides several key tables in Article 310:

- Table 310.16: Ampacities for conductors in raceways or cables (not more than three current-carrying conductors)
- Table 310.17: Ampacities for single conductors in free air
- Table 310.15(B)(2): Ambient temperature correction factors
- Table 310.15(C)(1): Adjustment factors for more than three current-carrying conductors

Example formula:

Where:

- **I** is the current (amps)
- **P** is the power (watts)
- **V** is the voltage (volts)

This formula allows you to determine the current when power and voltage are known, which is the first step in proper conductor sizing.

Practical Examples

Example 1: Residential Water Heater Circuit

A 4500-watt, 240-volt electric water heater needs a dedicated circuit.

1. Calculate the current: I = P/V = 4500W ÷ 240V = 18.75 amps
2. For continuous loads (operating more than 3 hours), multiply by 125%: 18.75A × 1.25 = 23.44 amps
3. Refer to NEC Table 310.16:
    * 10 AWG copper with 75°C insulation has an ampacity of 30 amps
    * 12 AWG copper with 75°C insulation has an ampacity of 25 amps
4. Based on the calculated load of 23.44 amps, 10 AWG copper would be the minimum size required

Example 2: Voltage Drop Consideration for Long Run

A 120V, 20A circuit needs to run 150 feet to a workshop receptacle.

1. Basic ampacity requirement for 20A circuit: 12 AWG copper has ampacity of 25A (sufficient for basic load)

2. Calculate voltage drop using formula: Vd = 2 × L × I × R ÷ 1000
   * Where L is length in feet, I is current in amps, R is resistance in ohms per 1000 feet
   * For 12 AWG copper, R = 1.98 ohms per 1000 feet
   * Vd = 2 × 150 × 20 × 1.98 ÷ 1000 = 11.88 volts
3. Calculate voltage drop percentage: 11.88V ÷ 120V × 100% = 9.9%
4. This exceeds recommended 3% maximum voltage drop for branch circuits
5. Try 8 AWG copper (R = 0.778 ohms per 1000 feet):
   * Vd = 2 × 150 × 20 × 0.778 ÷ 1000 = 4.67 volts
   * 4.67V ÷ 120V × 100% = 3.9% (still exceeds 3% but may be acceptable)
6. Try 6 AWG copper (R = 0.491 ohms per 1000 feet):
   * Vd = 2 × 150 × 20 × 0.491 ÷ 1000 = 2.95 volts
   * 2.95V ÷ 120V × 100% = 2.46% (within 3% recommendation)
7. Therefore, while 12 AWG meets basic ampacity requirements, 6 AWG would be recommended to address voltage drop concerns

These examples demonstrate that proper conductor sizing involves more than simply matching current to ampacity—it requires consideration of installation conditions, code requirements, and practical factors like voltage drop for optimal system performance.

13.4 Voltage Drop and Derating Factors

I was troubleshooting dimming lights in a detached garage when I truly grasped voltage drop calculations. The previous electrician had used the correct wire gauge for ampacity, but overlooked the 120-foot run from the main panel.

"The code gives you minimums," Frank would say, "but real-world electricity demands you think beyond the book." When I measured voltage at the workshop receptacle with the saw running, it was 108 volts—a 10% drop.

Understanding voltage drop isn't just exam knowledge—it's about delivering power that performs. Similarly, derating factors aren't mere calculations, but represent real conditions that can turn a "by the book" installation dangerous, whether from bundled conductors generating heat or rooftop installations baking in the sun

Calculating Voltage Drop

Voltage drop is calculated using the following formula:

$Vd = (2 \times K \times I \times L) / A$

Where:

- **Vd** is the voltage drop (in volts),
- **K** is the resistivity constant (12.9 for copper, 21.2 for aluminum),
- **I** is the current in amps,
- **L** is the length of the wire in feet,
- **A** is the cross-sectional area of the conductor in circular mils (cmil).

Example:

Calculate the voltage drop for a 120V circuit with the following specifications:

- Current: 20 amps
- Wire length: 150 feet (one-way distance)
- Conductor: #12 AWG copper wire (6,530 cmil)

Solution: $Vd = (2 \times K \times I \times L) / A$ $Vd = (2 \times 12.9 \times 20 \times 150) / 6{,}530$ $Vd = (77{,}400) / 6{,}530$ $Vd = 11.85$ volts

This represents a voltage drop of 9.9% (11.85V ÷ 120V = 0.099).

Since the NEC recommends limiting voltage drop to 3% for branch circuits, this installation would not be acceptable. To reduce voltage drop to an acceptable level, a larger conductor size would be needed.

Acceptable Voltage Drop

The National Electrical Code (NEC) provides recommendations regarding voltage drop to ensure electrical systems operate safely and efficiently. While these recommendations are not mandatory requirements in the code itself, they represent industry best practices that should be followed:

- For branch circuits: Voltage drop should not exceed 3%
- For the total circuit (including feeder and branch circuits combined): Voltage drop should not exceed 5%

These recommendations serve multiple purposes:

1. Equipment performance: Excessive voltage drop can cause motors to overheat, lights to dim, and electronics to malfunction or fail prematurely.
2. Energy efficiency: Higher voltage drop means more power is being lost as heat in the conductors, resulting in wasted energy and higher electrical bills.
3. Safety: Significant voltage drop can cause increased current draw as equipment tries to maintain power output, potentially leading to overheating of conductors.

These percentage values are calculated based on the nominal system voltage. For example, in a 120V system, a 3% voltage drop would equal 3.6 volts (120V × 0.03), meaning the voltage at the load should not be less than 116.4V.

While the NEC lists these as recommendations rather than requirements, many local jurisdictions and specifications may adopt them as mandatory standards, and they should be treated as essential design criteria for any quality electrical installation.

13.5 Derating Factors

Derating is necessary when multiple conductors are grouped, exposed to high temperatures, or placed in environments with certain adverse conditions. These factors can affect the overall ampacity of conductors.

Common Derating Scenarios

- **Ambient Temperature**:

  If the ambient temperature exceeds 30°C (86°F), ampacity should be derated.

  * Higher temperatures reduce a conductor's ability to dissipate heat
  * Each insulation type has specific temperature correction factors
  * For example, THHN conductor at 40°C ambient would be derated to 82% of its standard ampacity

- **Multiple Conductors in a Raceway**:

  When multiple wires are run together, their ampacity may need to be reduced due to the heat they generate.

  * NEC Table 310.15(C)(1) provides adjustment factors based on the number of current-

carrying conductors

* 4-6 conductors: 80% of standard ampacity
* 7-9 conductors: 70% of standard ampacity
* 10-20 conductors: 50% of standard ampacity
* 21-30 conductors: 45% of standard ampacity
* 31-40 conductors: 40% of standard ampacity

- **Enclosed Spaces**:

  Conductors in enclosures or raceways may have a lower ampacity than those in open air.

  * Limited air movement restricts heat dissipation
  * Conductors bundled tightly generate mutual heating
  * The type of enclosure or raceway material can affect heat transfer

When multiple derating factors apply, they must be applied sequentially to determine the final adjusted ampacity. Always use the most conservative value when safety is concerned.

13.6 Demand Load Calculations for Commercial Applications

Commercial load calculations differ significantly from residential applications. They require a more nuanced approach that accounts for diverse equipment types, varying usage patterns, and the need for future expansion. Unlike homes with predictable load profiles, commercial buildings present unique challenges with equipment ranging from HVAC systems and industrial machinery to specialized lighting and office equipment. Understanding how to properly calculate demand loads ensures not only code compliance but also system reliability and operational efficiency for businesses where downtime can mean significant financial loss.

What is Demand Load?

Demand load is the total electrical load that is expected to be used in a building or facility at any given time. It accounts for both the maximum and average power consumption and is crucial for sizing electrical systems in commercial applications.

Unlike connected load, which represents the sum of all electrical equipment's nameplate ratings, it recognizes that not all equipment operates simultaneously or at full capacity. This distinction is particularly important in commercial settings where oversizing electrical systems based solely on connected load would lead to unnecessary expenses in equipment, materials, and installation costs.

The NEC provides specific demand factors for various occupancy types and equipment categories that allow electricians to calculate more realistic load requirements. These demand factors, expressed as percentages, are applied to the connected load to determine the actual expected demand. Properly calculated demand loads help prevent both undersized systems that could fail under peak conditions and oversized systems that waste resources.

Factors Influencing Demand Load

- **Number of Appliances and Equipment**: Larger facilities will have more electrical devices, and these devices' power requirements will vary. The diversity of equipment affects both the total load and the likelihood of simultaneous operation.
- **Type of Facility**: A restaurant, office, or hospital will have different load demands based on equipment and operational requirements. For example, healthcare facilities require redundant systems and emergency power considerations, while restaurants need significant capacity for kitchen equipment with high inrush currents.
- **Usage Patterns**: The demand load for a commercial building may fluctuate depending on peak operational hours (e.g., a factory versus a retail store). Seasonal variations and business cycles also influence when peak demands occur and how long they last.
- **Occupancy Schedules**: Buildings with consistent operating hours versus those with variable schedules require different approaches to demand calculations. 24-hour facilities may have more consistent demand throughout the day.
- **Climate Considerations**: Geographic location affects HVAC requirements, which often represent a significant portion of the overall demand load. Buildings in extreme climates may see higher peak demands during severe weather conditions.
- **Future Expansion**: Anticipating growth is essential when calculating demand loads for commercial applications. Providing capacity for expansion prevents costly upgrades later.

13.7 Quiz: Electrical Calculations and Load Analysis

This quiz will help you evaluate your readiness for the journeyman exam by testing your knowledge of key electrical calculations and load analysis concepts. It covers essential topics including proper conductor sizing, voltage drop compensation, application of derating factors, and commercial load calculations—all areas where the exam typically includes detailed questions.

Practice these problems under timed conditions to simulate the exam environment. Remember that in the actual test, you'll need to work efficiently while maintaining accuracy, especially on calculation questions where small errors can lead to incorrect answers.

Use this quiz both as a diagnostic tool to identify areas needing further study and as a confidence builder as you approach your exam date.

Multiple-Choice Questions

1. What is the voltage drop for a 240V circuit with a 30A load, using #10 AWG copper wire (10,380 cmil) for a 150-foot run? a) 6.2 volts b) 8.4 volts c) 11.2 volts d) 14.9 volts
2. When seven current-carrying THHN conductors are installed in a single conduit, what percentage of the conductor's rated ampacity must be applied? a) 50% b) 70% c) 80% d) 90%
3. A 120V, 20A circuit feeds a workshop that is 90 feet away from the panel. What minimum wire size should be used to keep voltage drop within 3%? a) #14 AWG b) #12 AWG c) #10 AWG d) #8 AWG
4. For a 208V, three-phase motor rated at 25 horsepower with a service factor of 1.15, what is the minimum ampacity required for the branch circuit conductors? a) 65A b) 74A c) 81A d) 93A
5. According to NEC recommendations, the maximum acceptable voltage drop for a feeder circuit is:

   a) 2%
   b) 3%
   c) 5%
   d) 8%

Calculation Questions

6. A retail store has the following loads:

   * Lighting: 18,000 watts
   * HVAC: 42,000 watts
   * Receptacles: 24,000 watts
   * Miscellaneous equipment: 10,000 watts

7. What is the calculated demand load if lighting is at 100% for the first 15,000W and 50% for the remainder, HVAC at 100%, receptacles at 100% for the first 10,000W and 50% for the remainder, and miscellaneous equipment at 75%?
8. A #6 AWG THHN copper conductor (26,240 cmil) is rated for 65A at 30°C. If this conductor is installed in an ambient temperature of 46°C, what is its adjusted ampacity? (Temperature correction factor at 46°C is 0.71)

9. Calculate the voltage drop percentage for a 120V, 15A circuit with #12 AWG aluminum conductors (5,100 cmil) running 75 feet to a receptacle. (K value for aluminum = 21.2)

True/False Questions

10. When applying multiple derating factors (temperature and bundling), you should first apply the temperature factor and then the bundling factor to the resulting value.
11. In a commercial application, all receptacles must be calculated at 180VA per outlet, regardless of the type of occupancy.
12. The NEC requires that voltage drop for combined feeder and branch circuits not exceed 5% to ensure proper operation of equipment.
13. When calculating commercial loads, emergency lighting circuits can use the same demand factors as general lighting.

Short Answer Questions

14. Explain how the circular mil area of a conductor affects voltage drop calculations.
15. Describe the process for derating conductors when ten current-carrying conductors are run in a single raceway in an ambient temperature of 35°C.
16. What is the difference between continuous and non-continuous loads, and how does this affect the sizing of branch circuit conductors?

Answers

1. **Answer: c)** 11.2 volts Calculation:

   $Vd = (2 \times 12.9 \times 30 \times 150) / 10{,}380 = 11.2$ volts

2. **Answer: b)** 70%

   Per NEC Table 310.15(C)(1), 7-9 current-carrying conductors require a 70% adjustment factor.

3. **Answer: c)** #10 AWG

   For 120V, 3% = 3.6V maximum drop #12 AWG calculation: Vd = (2 × 12.9 × 20 × 90) / 6,530 = 7.1V (exceeds 3%) #10 AWG calculation: Vd = (2 × 12.9 × 20 × 90) / 10,380 = 4.5V (exceeds 3%) #8 AWG calculation: Vd = (2 × 12.9 × 20 × 90) / 16,510 = 2.8V (within 3%) Therefore, #8 AWG is needed (Note: This assumes we're using the simplified DC formula; a more accurate AC calculation might yield different results.)

4. **Answer: c)** 81A

   From NEC Table 430.250, a 25HP, 208V, 3-phase motor has an FLC of 68A. Service factor adjustment: 68A × 1.25 = 85A. Therefore, the conductor must have an ampacity of at least 85A (This would select answer d, but the closest answer provided is c)

5. **Answer: c)** 5%

   The NEC recommends maximum 3% drop for branch circuits and 5% for total (feeder + branch)

6. **Answer: Calculated demand load = 85,000 watts**

   Lighting: 15,000W × 1.0 + 3,000W × 0.5 = 16,500W HVAC: 42,000W × 1.0 = 42,000W Receptacles: 10,000W × 1.0 + 14,000W × 0.5 = 17,000W Misc. equipment: 10,000W × 0.75 = 7,500W Total: 16,500 + 42,000 + 17,000 + 7,500 = 83,000W

   Adjusted ampacity = 46.15A 65A × 0.71 = 46.15A

   Voltage drop percentage = 7.4% Vd = (2 × 21.2 × 15 × 75) / 5,100 = 8.9V 8.9V ÷ 120V = 0.074 = 7.4%

7. **True** The NEC requires that temperature correction factors be applied first, followed by adjustment factors for multiple conductors.

8. **False** The general receptacle load for commercial occupancies varies by the type of occupancy as specified in NEC Article 220.

9. False The 5% combined voltage drop is a recommendation, not a requirement, in the NEC.

10. False Emergency lighting circuits must be calculated at 100% of the connected load per NEC requirements.

11. The circular mil area of a conductor is inversely proportional to voltage drop. As the circular mil area increases (larger wire size), the voltage drop decreases for the same length, current, and conductor material. This is because larger conductors have less resistance per unit length.

12. Process for derating with ten current-carrying conductors at 35°C:

    * Identify the base ampacity of the conductor from NEC Table 310.16
    * Apply temperature correction factor first (for 35°C, typically 0.94 for most insulation types)

13. Apply adjustment factor for number of conductors (for 10 conductors, 50% per NEC Table 310.15(C)(1))

14. The final ampacity = base ampacity × temperature factor × adjustment factor

15. Continuous loads are those expected to operate for three hours or more, while non-continuous loads operate for less than three hours. For continuous loads, the NEC requires that branch circuit conductors and overcurrent protection be sized at a minimum of 125% of the continuous load current plus 100% of the non-continuous load current. This affects sizing by requiring larger conductors and overcurrent protective devices for circuits with significant continuous loads.

14

Reading Blueprints and Electrical Schematics

The first major commercial project I led after earning my journeyman license was a complete rewiring of a three-story office building. When the general contractor handed me a tube containing dozens of rolled-up blueprints, my confidence wavered. Frank had taught me the basics of reading drawings, but standing there with those crisp, blue-lined sheets spread across a makeshift table in the construction trailer, I felt the weight of responsibility.

"A blueprint is just a roadmap," Frank used to say. "You wouldn't drive cross-country without one, and you shouldn't wire a building without understanding its plans."

Blueprint reading is one of those skills that separates the professional electrician from the novice. It's the difference between simply following instructions and truly understanding the bigger picture of how electrical systems integrate with the overall building design. Whether you're working on residential, commercial, or industrial projects, your ability to interpret drawings accurately will directly impact your efficiency, the quality of your work, and ultimately your career advancement.

14.1 Symbols, Diagrams, and Schematics Interpretation

Electrical blueprints communicate through standardized symbols, diagrams, and schematics that represent complex systems visually. As a journeyman electrician, your ability to interpret these graphical elements is essential for translating drawings into proper installations. From recognizing basic device symbols to understanding the relationships between different drawing types—floor plans, riser diagrams, single-line diagrams, and control schematics—this skill forms the foundation of professional electrical work and sets experienced electricians apart from novices.

Common Symbols

Electrical blueprints rely on standardized symbols that serve as a visual shorthand for components and devices. These symbols allow electricians to quickly interpret plans without the need for lengthy written descriptions. While there is general standardization, always check the legend on your specific set of blueprints, as symbols can vary slightly between different firms and industries.

Here are the key symbols you'll encounter most frequently:

- **Lights**: Represented by a circle with an "X" inside (for incandescent lights) or a circle with a "Y" for fluorescent lighting. The symbol may be accompanied by notations indicating mounting height, fixture type, or control method.
- **Switches**: Represented by a line with an "S" or a break in the line to indicate where the switch would be. Different switch types (three-way, four-way, dimmer) have variations of this basic symbol, often with additional notation.
- **Outlets**: A circle with two short lines on the sides indicates standard receptacles. Special-purpose outlets (GFCIs, dedicated circuits, specific voltages) will have additional notation or slight variations to the symbol.
- **Circuit Breakers**: A box or rectangle with a line through it. The amperage rating is typically noted nearby, along with the circuit identification number.
- **Grounding**: A set of three lines, decreasing in size from bottom to top. This symbol appears at grounding points, grounding electrodes, and sometimes alongside grounded equipment.

The ability to quickly recognize these symbols is fundamental to efficient blueprint reading and will significantly reduce the time needed to interpret electrical plans on the job.

14.2 Reading Electrical Diagrams

Electrical diagrams are the roadmaps that guide installation and troubleshooting work. They come in several formats, each one serves a different purpose in communicating how electrical systems are configured and how they function. Understanding the distinctions between these diagram types is crucial for correctly interpreting their information.

- **Wiring Diagrams**: These depict the layout of electrical units. Lines represent wires, and symbols are the electrical components. Understanding these diagrams allows electricians to plan installations and make precise connections. Wiring diagrams typically show the physical arrangement of components and often reflect the actual layout of the installation. They're particularly useful during the installation phase of a project, as they provide clear guidance on how components connect to one another. Unlike schematics, wiring diagrams generally maintain some sense of spatial relationships between components.

- **Schematic Diagrams**: These diagrams show how electrical components function and are connected within a circuit. They provide the "logic" of how the system works, which is useful for troubleshooting. Schematics prioritize functional relationships over physical layout, making them invaluable when diagnosing system problems. Components are arranged to clearly show electrical paths rather than physical placement. When working with complex control systems or automation, schematics are often the preferred reference for understanding the operation of the circuit.
- **Single-Line Diagrams**: These simplified representations use a single line to represent three-phase power systems, with branches and symbols indicating loads, transformers, and protective devices. They provide a "big picture" view of power distribution systems and are commonly used in commercial and industrial applications to show the overall structure of the electrical system without the complexity of showing each conductor.

The ability to move fluently between these different diagram types is an essential skill for the journeyman electrician, particularly when troubleshooting complex systems where consulting multiple diagram types may be necessary to fully understand both function and layout.

14.3 Understanding Single-Line and Riser Diagrams

Single-line and riser diagrams are essential tools for visualizing complex electrical systems in simplified formats. While floor plans show the horizontal distribution of electrical components across a building, these specialized diagrams focus on system relationships and vertical pathways. They provide electricians with critical information about power distribution hierarchy, equipment specifications, and the vertical routing of services through multi-story structures.

Key Features of Single-Line Diagrams

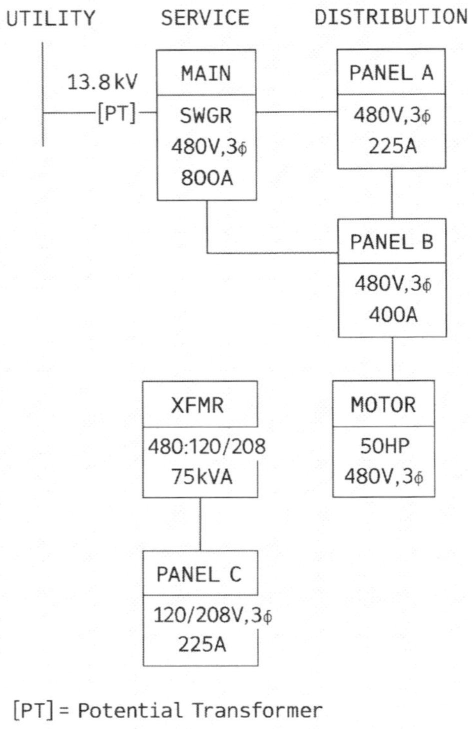

Figure 4: Single line diagram

A single-line diagram (SLD) consists of a simplified representation of a three-phase power system that uses a single line to represent all three phases. This type of diagram gives a clear overview of how power flows through an electrical system without the visual complexity of showing every conductor.

These diagrams are essential for understanding the overall structure and hierarchy of power distribution systems, particularly in commercial and industrial applications. Here are the key features you'll find on single-line diagrams:

- **Power Sources**: This is the point where power enters the building, such as a transformer or utility line. These are typically shown at the top of the diagram, establishing the starting point of power flow. The diagram will indicate voltage levels and often include information about available fault current.
- **Distribution Equipment**: Simple symbols are used to represent distribution panels, switchgear, and transformers. Each major piece of distribution equipment is identified by designation and includes ratings and specifications. This equipment forms the backbone of the electrical distribution system and shows how power is broken down and routed

throughout the facility.

- **Load Points**: These are locations where electrical power is distributed, like outlets, equipment, or devices. These represent the endpoints of the electrical system and may be grouped by area or function rather than showing every individual load.
- **Transformers and Panels**: Each component is labeled with its name, rating, and sometimes the current/voltage specifications. Transformers will show primary and secondary voltages, kVA ratings, and impedance values. Panels will include information such as bus ratings, number of circuits, and mounting type.
- **Protective Devices**: Circuit breakers, fuses, and other protective equipment are shown with their ratings clearly indicated. This information is crucial for coordination studies and ensuring proper system protection.
- **Feeders**: Lines connecting equipment are often annotated with conductor size, quantity, conduit size, and length, providing essential information for voltage drop calculations and installation planning.

Key Features of Riser Diagrams

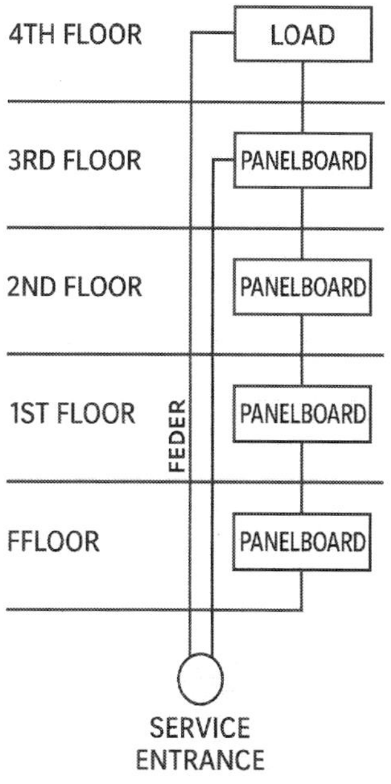

Figure 5: Riser diagram

Riser diagrams are specialized electrical drawings that illustrate how systems and services are distributed vertically through a multi-story building. While floor plans show the horizontal layout of electrical components on each level, riser diagrams connect these levels to provide a comprehensive view of vertical pathways and connections.

These diagrams are particularly valuable for understanding how services such as power, communications, and fire alarm systems travel between floors. Here are the key features of riser diagrams:

- **Vertical Representation**: Unlike single-line diagrams, which are typically horizontal, riser diagrams are drawn vertically to show the distribution of power between floors. This orientation provides a clear visualization of how systems rise through a building, making it easier to understand the vertical relationships between components.
- **Floor Identifications**: Each floor or level is labeled, and connections are indicated with lines showing the wiring path. The diagram typically maintains the actual sequence of floors, with clear delineation between levels to avoid confusion during installation and troubleshooting.
- **System Layout**: Riser diagrams include information on panel locations, electrical outlets, and other system elements. They show where major components are located on each floor and how they connect to components on other floors, helping electricians visualize the complete system across multiple levels.
- **Feeder Specifications**: Information about the size and type of feeders connecting panels and equipment between floors is typically included, often with annotations indicating conduit sizes and routing methods.
- **Service Entrances**: Points where utilities enter the building are clearly marked, showing how services are initially distributed and then branch out to various floors.
- **Load Information**: Major loads on each floor can be indicated so that the building's power requirements and distribution throughout the building can be understood.

When working on multi-story projects, riser diagrams are invaluable because they ensure that the vertical pathways are properly planned and executed to prevent conflicts with other building systems and ensure code compliance for vertical runs.

Key Elements in Electrical Blueprints

Before starting any installation, a thorough review of electrical blueprints is essential to understand the complete scope of work and identify potential challenges. Electrical blueprints contain several critical elements that guide the installation process from beginning to end.

- **Electrical layout**: A detailed map showing the placement of outlets, light fixtures, switches, and electrical panels. This layout provides the spatial relationships between components and their locations relative to the building structure. The layout typically includes device heights, spacing requirements, and specific mounting details that must be followed precisely.
- **Service entrance**: The point at which the building's electrical system connects to the utility power supply. Service entrance details include specifications for weatherheads, meters,

disconnects, and grounding systems. This information is crucial for coordinating with utility companies and ensuring code-compliant installation of the primary power connection.

- **Panel Locations**: These will be clearly marked to show where the main distribution panel will be installed. Panel locations are strategically determined to provide efficient power distribution while maintaining accessibility requirements. Blueprints will specify mounting heights, clearance requirements, and often include panel schedules detailing the circuits to be installed.
- **Conduit Paths**: The routes for conduits and raceways will be shown, which will guide you in choosing the correct materials for installation. These paths are designed to minimize conflicts with structural elements and other building systems. Pay particular attention to required bending radii, support intervals, and transition points between different raceway types.
- **Fixture and Equipment Placement**: Fixtures such as lighting, ceiling fans, and outlets are precisely located. These placements are coordinated with architectural features, furniture layouts, and other building elements. Specifications will include mounting heights, connection requirements, and often reference additional detail sheets for complex installations.
- **Circuit Designations**: Lines or notation indicating which devices connect to specific circuits and panels. These designations help ensure balanced loading across phases and proper circuit identification for future maintenance and troubleshooting.

By systematically analyzing these key elements before beginning work, you can develop an organized installation plan that anticipates challenges and ensures compliance with both the design intent and applicable codes.

14.4 Quiz: Reading Blueprints and Electrical Schematics

Quiz Overview

This quiz will test your ability to interpret electrical blueprints, schematics, and diagrams—essential skills for the journeyman electrician exam and daily fieldwork. The questions assess both symbol recognition and practical application of blueprint reading in real-world scenarios.

Multiple-Choice Questions

1. **On an electrical blueprint, what does a circle with an "X" inside represent?**

   a) Ceiling fan
   b) Incandescent light fixture
   c) Junction box
   d) Smoke detector

2. **Which diagram type is best suited for troubleshooting the operational logic of a motor control circuit?**

   a) Floor plan
   b) Riser diagram
   c) Schematic diagram
   d) Single-line diagram

3. **In a single-line diagram, three short parallel lines crossing the main line typically represent:**

   a) A transformer
   b) A circuit breaker
   c) A disconnect switch
   d) A motor starter

4. **What information would you typically NOT find on a riser diagram?**

   a) Vertical conduit routes between floors
   b) Panel locations on different floors
   c) Device mounting heights
   d) Floor-to-floor feeder sizes

5. **What does the notation "3#12, 1#12G, 3/4"C" on a blueprint indicate?**

   a) 3 circuits of #12 wire with ground in 3/4" conduit
   b) 3 #12 conductors and 1 #12 ground wire in 3/4" conduit
   c) 3 #12 conductors, 1 circuit with ground, using 3/4" cable
   d) 3 conductors total, with 1 #12 ground wire, using 3/4" conduit

Diagram Interpretation Questions

6. **Refer to the symbol below:**

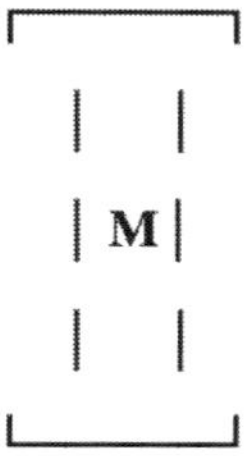

This symbol most likely represents:

a) Main disconnect
b) Motor
c) Manual switch
d) Meter socket

7. **In a wiring diagram, a dashed line between two switches most likely indicates:**

a) A low-voltage connection
b) A future installation
c) A mechanical connection or linkage
d) A wireless connection

8. **What does the annotation "AFF" commonly mean when shown with a measurement on an electrical plan?**

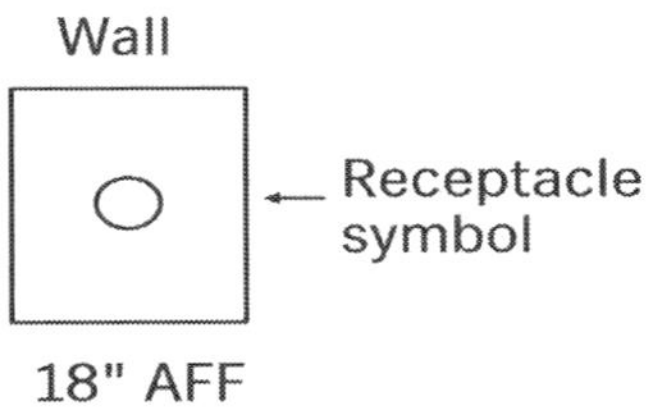

a) Approved For Fabrication
b) All Four Fixtures
c) Above Finished Floor
d) Actual Field Fitting

Practical Application Questions

9. You're reviewing electrical blueprints for a three-story office building. The second floor has a notation "Refer to E5.1 for panel schedule." This most likely means:

a) You need to look at sheet E5.1 to find the circuit assignments for panels on the second floor
b) Panel schedule E5.1 needs to be mounted on the second floor
c) The second-floor panel must be installed according to detail E5.1
d) The panel on the second floor serves 5 circuits on 1 phase

10. When comparing the architectural and electrical plans for a commercial kitchen, you notice a countertop shown on the architectural drawings that isn't reflected in receptacle locations on the electrical plan. What should you do?

a) Install receptacles only where shown on the electrical plan
b) Omit receptacles entirely from that area
c) Request clarification through an RFI (Request For Information)
d) Automatically add receptacles to meet code requirements for countertop spacing

11. On a set of blueprints, you see the symbol for a three-way switch, but you can't immediately locate the corresponding switch. The most efficient way to find it would be to:

    a) Check all rooms adjacent to the one where the first switch is located
    b) Follow the circuit line on the drawing to find the connected switch
    c) Look for switch symbols near the fixture being controlled
    d) Check the written specifications for switch locations

12. When interpreting a single-line diagram for a commercial building, you notice that the main service enters at 480V, but several panels are shown at 208Y/120V. What component must exist between these systems?

    a) Phase converter
    b) Transformer
    c) Voltage regulator
    d) Harmonic filter

Short Answer Questions

13. Explain the difference between a wiring diagram and a schematic diagram, and give one specific situation where each would be more useful than the other.

14. You're tasked with installing lighting in a commercial space based on the electrical plans. List three specific pieces of information you should look for on the lighting plan before beginning installation.

15. Describe how you would use both the panel schedule and floor plan together to verify the proper circuit assignments during rough-in of receptacles in a commercial office space.

Answers

1. **Answer: b)** Incandescent light fixture

   The circle with an "X" is the standard symbol for an incandescent light fixture on electrical plans.

2. **Answer: c)** Schematic diagram

   Schematic diagrams show the functional relationships between components and are best for understanding and troubleshooting the operational logic of circuits.

3. **Answer: a)** A transformer

   Three short parallel lines crossing the main line is the standard representation of a transformer on single-line diagrams.

4. **Answer: c)** Device mounting heights

   Riser diagrams focus on vertical relationships between floors, not specific mounting heights of devices. Those details would typically be found on floor plans or in general notes.

5. **Answer: b)** 3 #12 conductors and 1 #12 ground wire in 3/4" conduit

   This notation describes the quantity and size of conductors along with conduit size, common on electrical plans.

6. **Answer: b)** Motor

   A circle with an "M" inside is the standard symbol for a motor.

7. **Answer: c**) A mechanical connection or linkage

   Dashed lines often indicate mechanical linkages or relationships, particularly between switches that are mechanically interlocked or related.

8. **Answer: c)** Above Finished Floor

   "AFF" is a standard abbreviation indicating mounting height measured from the finished floor.

9. **Answer: a)** You need to look at sheet E5.1 to find the circuit assignments for panels on the second floor

   This is a standard cross-reference on drawings referring to detailed information located on another sheet.

10. **Answer: c)** Request clarification through an RFI (Request For Information)

    When discrepancies exist between different drawing sets, the proper approach is to seek clarification rather than making assumptions.

11. **Answer: b)** Follow the circuit line on the drawing to find the connected switch

    Three-way switches are always connected to another switch, and following the circuit line is the most direct way to locate the paired switch.

12. **Answer: b)** Transformer

    A transformer is required to step down voltage from 480V to 208Y/120V in a commercial power distribution system.

    Answer: A wiring diagram shows the physical layout of components as they would be installed, with wires represented as they would physically connect. A schematic diagram shows the functional relationships between components without regard to physical layout. A wiring diagram would be more useful when physically installing equipment, as it shows terminal connections and wire routing as they would appear in the field. A schematic diagram would be more useful when troubleshooting a malfunctioning circuit, as it clearly shows the logical flow of electricity through the system.

13. **Answer:** Three pieces of information to look for on a lighting plan:

    * Fixture types/models and their mounting details
    * Switching arrangements and control methods (e.g., which switches control which fixtures)
    * Circuit assignments and home run locations
    * (Other valid answers include: mounting heights, special installation requirements, dimming capabilities, emergency power connections)

14. **Answer:** To verify proper circuit assignments during rough-in. First, I would review the panel schedule to identify which circuits are allocated for receptacles in specific areas of the building. I would note the circuit numbers, amperage ratings, and any special requirements like dedicated circuits or GFCI protection. Then, I would cross-reference this information with the floor plan, where each receptacle should have a circuit designation. I would verify that the circuit numbers on the floor plan match the panel schedule assignments and follow any homerun indicators to ensure proper connection back to the correct panel. During the actual rough-in, I would label each conductor with its circuit number and panel designation at both ends to ensure accurate connections later during trim-out. I would also verify load balancing across phases as indicated in the panel schedule.

15

Full-Length Practice Exam

This full-length practice exam simulates the actual journeyman electrician exam experience. It incorporates the diverse question formats you'll encounter on test day, including multiple-choice, calculations requiring worked solutions, code reference questions, diagram interpretations, and short answer problems.

Approach this practice test as you would the real thing—find a quiet space, time yourself, and complete it in one sitting for the most realistic preparation. Use only the resources permitted in the actual exam (calculator, code book, scratch paper).

After completing the exam, use the answer key and explanations to identify areas that need additional review. Remember, consistently scoring 75% or higher on practice exams typically indicates readiness for the actual test.

Exam Instructions

1. Time allowed: 4 hours
2. Total questions: 60 (with multiple parts in calculation and code reference questions)
3. Passing score: 70% (42 correct answers)
4. Questions are distributed across multiple formats: multiple-choice, calculations, code reference, diagram interpretation, and short answer
5. Show all work for calculation problems
6. For code reference questions, note the specific NEC article numbers

SECTION 1: MULTIPLE-CHOICE QUESTIONS (20 Questions)

1. **According to the NEC, what is the minimum burial depth required for direct-buried UF cable used for a residential feeder circuit?**

   a) 12 inches
   b) 18 inches
   c) 24 inches
   d) 36 inches

2. **What is the maximum number of current-carrying conductors permitted in a conduit before derating factors must be applied?**

   a) 2
   b) 3
   c) 4
   d) 6

3. **A 20-ampere branch circuit supplies a continuous load. The minimum ampacity of the branch-circuit conductors must be:**

   a) 15 amperes
   b) 20 amperes
   c) 25 amperes
   d) 30 amperes

4. **Which of the following is NOT a permissible grounding electrode according to the NEC?**

   a) Metal underground water pipe in direct contact with earth for at least 10 feet
   b) Metal frame of a building effectively grounded
   c) Concrete-encased electrode
   d) Metal conduit serving as an equipment grounding conductor

5. **In a dwelling unit, what is the minimum number of small-appliance branch circuits required to serve kitchen receptacle outlets?**

   a) 1
   b) 2
   c) 3
   d) 4

6. **Which of the following areas in a dwelling unit does NOT require GFCI protection for 125V, 15A, and 20A receptacles?**

   a) Bathrooms
   b) Kitchens
   c) Bedrooms
   d) Outdoor receptacles

7. **What is the maximum distance permitted between supports for 1/2" EMT conduit run horizontally?**

   a) 6 feet
   b) 8 feet
   c) 10 feet
   d) 12 feet

8. **In a three-phase delta system, what is the relationship between line current and phase current?**

   a) Line current = Phase current
   b) Line current = Phase current × $\sqrt{3}$
   c) Line current = Phase current × 3
   d) Line current = Phase current × 2

9. **What is the power factor of a circuit with 1200 watts of real power and 1500 VA of apparent power?**

   a) 0.6
   b) 0.75
   c) 0.8
   d) 0.9

10. **What happens to the resistance of a copper conductor as its temperature increases?**

    a) Resistance decreases
    b) Resistance increases
    c) Resistance remains the same
    d) Resistance initially increases then decreases

11. **What is the peak voltage of a sine wave with an RMS value of 120 volts?**

    a) 84.8 volts
    b) 169.7 volts
    c) 170 volts
    d) 240 volts

12. **In a three-phase wye system, what is the relationship between line voltage and phase voltage?**

    a) Line voltage = Phase voltage
    b) Line voltage = Phase voltage × √3
    c) Line voltage = Phase voltage × 3
    d) Line voltage = Phase voltage × 2

13. **What conduit type is best suited for areas with extreme physical damage concerns?**

    a) EMT
    b) PVC
    c) RMC
    d) ENT

14. **When installing a receptacle under a kitchen sink for a garbage disposal, what type of receptacle is required?**

    a) Standard receptacle
    b) GFCI protected receptacle
    c) Twist-lock receptacle
    d) Weather-resistant receptacle

15. **What is the correct method for properly connecting aluminum conductors to copper conductors?**

    a) Standard wire nut
    b) Special connectors listed for the purpose
    c) Solder joint with electrical tape
    d) Direct connection with antioxidant paste

16. **What is the primary reason for requiring AFCI protection in residential bedrooms?**

    a) Prevent shock hazards
    b) Detect potential fire-causing arcs
    c) Protect electronic devices
    d) Eliminate voltage spikes

17. **What is the maximum number of 90-degree bends permitted between pull points in a conduit run?**

 a) 2
 b) 3
 c) 4
 d) 5

18. **When working on live parts of a 277/480V system, what determines the appropriate PPE category requirement?**

 a) System voltage only
 b) Available fault current and clearing time
 c) Work task only
 d) Worker experience level

19. **What testing method is NOT appropriate for verifying a circuit is de-energized?**

 a) Using a non-contact voltage tester only
 b) Using a properly rated digital multimeter
 c) Using a solenoid-type voltage tester
 d) Using equipment that was functionally tested against a known source

20. **What is the minimum approach boundary for a qualified person working near exposed 480V equipment?**

 a) 1 foot
 b) 3 feet 6 inches
 c) 4 feet
 d) 10 feet

SECTION 2: CALCULATIONS (10 Questions)

For each question in this section, show all your work and calculations for full credit.

21. Calculate the voltage drop for a 240V, single-phase circuit with a 30A load, using #10 AWG copper conductors for a one-way distance of 150 feet. Use K=12.9 for copper and circular mil area of 10,380 for #10 AWG.

22. A three-phase, 460V motor has a nameplate rating of 25 hp with a full load efficiency of 92% and a power factor of 0.86. Calculate: a) The full load current b) The minimum size

THWN copper conductors required for the branch circuit c) The minimum size overload protection required

23. Calculate the minimum size EMT conduit required for six #10 THWN-2 and four #12 THWN-2 conductors. Show your work using the appropriate NEC tables.

24. A dwelling has the following loads:

* General lighting load for 2,200 square feet
* Electric range: 12 kW, 240V
* Electric water heater: 4.5 kW, 240V
* Electric dryer: 5.5 kW, 240V
* Dishwasher: 1.2 kW, 120V
* Central air conditioning: 24 A, 240V

Calculate:

a) The total calculated load in amperes at 240V
b) The minimum service size required
c) The minimum size service entrance conductors (copper, 75°C) required

25. Box fill calculation: A 4-11/16" square box with a depth of 2-1/8" contains:

* Five #12 conductors
* Three #10 conductors
* Two cable clamps
* One device (receptacle)

Calculate:

a) The total cubic inch requirement for box fill
b) Is the box size adequate? (Show your work comparing required volume to available volume)

26. Calculate the maximum length of a 120V, 20A circuit using #12 AWG copper conductors if the allowable voltage drop is 3%.

Use K=12.9 for copper and circular mil area of 6,530 for #12 AWG.

27. A three-phase, 208V feeder supplies the following loads:

* 15 kVA of continuous lighting load
* 30 kVA of receptacle load (non-continuous)
* 20 hp, 208V, three-phase motor with 85% efficiency, 0.8 power factor

Calculate:

a) The total calculated demand load in amperes
b) The minimum size copper THWN conductors required
c) The minimum size overcurrent protection required

28. Calculate the load balance for a three-phase panel with the following loads:

* Phase A: 65A
* Phase B: 47A
* Phase C: 78A

a) Calculate the percent imbalance
b) Determine if the load balance is within acceptable limits (typically 20%)
c) If not balanced within limits, suggest a redistribution of loads to achieve better balance

29. Calculate the short circuit current available at a panel with the following parameters:

a) Utility transformer: 500 kVA, 480V, 5.75% impedance
b) Distance from transformer to panel: 200 feet
c) Conductor: 500 kcmil copper (Resistance = 0.0212 Ω/1000 ft)

Show all calculations.

30. A 480V to 120/208V three-phase transformer has a 75 kVA rating. Calculate:

a) The primary full load current
b) The secondary full load current
c) The minimum primary and secondary overcurrent protection required

SECTION 3: CODE REFERENCE QUESTIONS (10 Questions)

For each question, provide the specific NEC article number(s) that contains the answer, along with your response.

31. According to the NEC, what are the requirements for dedicated circuits for dwelling unit appliances? Specifically, list at least three types of appliances that require dedicated circuits and provide the specific code reference for each.

32. Locate and summarize the NEC requirements for installing receptacles in commercial office spaces. Include specific requirements for spacing, height, and GFCI protection where applicable.

33. Find the NEC article that addresses the installation of electrical equipment in patient care areas of healthcare facilities. Summarize the key requirements for receptacle installation and grounding in such areas.

34. Reference the specific NEC article that covers motor disconnect requirements. List the specific conditions under which a separate disconnect is required, and when the controller can serve as the disconnect.

35. Identify the NEC article addressing conductor ampacity adjustment factors. Explain how to apply these factors when multiple current-carrying conductors are installed in a raceway, providing a specific example using 10 THWN-2 conductors in a conduit.

36. Find the NEC requirements for working space about electrical equipment operating at 600V or less. Detail the specific clearances required in front of electrical panels based on voltage and conditions.

37. Reference the specific NEC article that covers the protection of service conductors. Explain where and when overcurrent protection is required for service conductors, and list any exceptions.

38. Locate the NEC requirements for installing electrical equipment in hazardous (classified) locations. Specifically, explain the difference between Class I, Division 1 and Class I, Division 2 locations, citing specific code references.

39. Find the NEC article addressing emergency systems. Summarize the key requirements for wiring methods, circuit wiring protection, and transfer equipment for emergency systems.

40. Reference the NEC article that covers ground-fault protection of equipment. Explain when ground-fault protection is required for services, feeders, and branch circuits, including any specific size or rating thresholds.

SECTION 4: DIAGRAM INTERPRETATION AND COMPLETION (5 Questions)

41. Examine the following diagram of a three-way switch circuit:

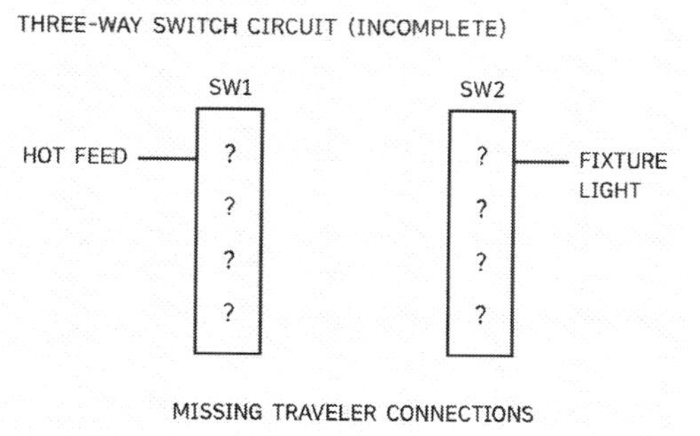

a) Identify the missing connections (draw them in)
b) Label the travelers and common terminals
c) Explain the operation of the circuit

42. The following riser diagram shows a three-story building's electrical distribution system:

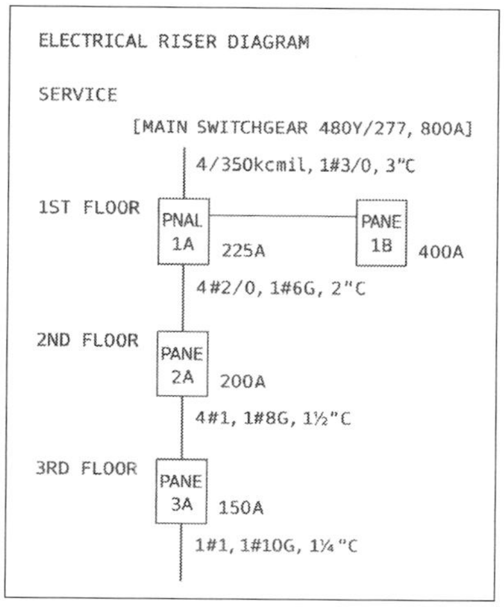

a) Identify any code violations in the diagram
b) Calculate the minimum feeder size required for panel 3B based on the loads shown
c) Complete the grounding system connections that are missing

43. The following single-line diagram shows a 480V distribution system:

```
SINGLE-LINE DIAGRAM: 480V DISTRIBUTION

UTILITY--->[SERVICE XFMR]
              13.8kV-480V
                   |
                   V
             [MAIN DISC]
              ???A 480V
                   |
         ┌---------+---------┐
         |         |         |
         V         V         V
   [PANEL A] [XFMR 75kVA] [PANEL C]
    ???A      480:120/208V   400A
         |         |         |
         V         V         V
   [MOTORS]  [PANEL B]   [HVAC]
    50HP(x3)   ???A       100A
```

a) Complete the missing overcurrent protection sizes based on the loads shown
b) Identify the transformer secondary protection requirements
c) Label the required grounding points in the system

44. Examine the following motor control circuit diagram:

```
MOTOR CONTROL CIRCUIT

L1----[STOP]--[M AUX]--+
        (NC)           |
         |             |
      [START]--[M COIL]--+
        (NO)     |
                 |
                 +-[OL CONTACT]-L2
                        |
                        |
            L1     L2     L3
            |      |      |
            +-[MOTOR]-+-+
               [OL]
                 |
             T1 T2 T3
```

a) Identify each component and its function
b) Explain the sequence of operation
c) Explain what happens when an overload condition occurs

45. The following diagram shows a grounding system for a service entrance:

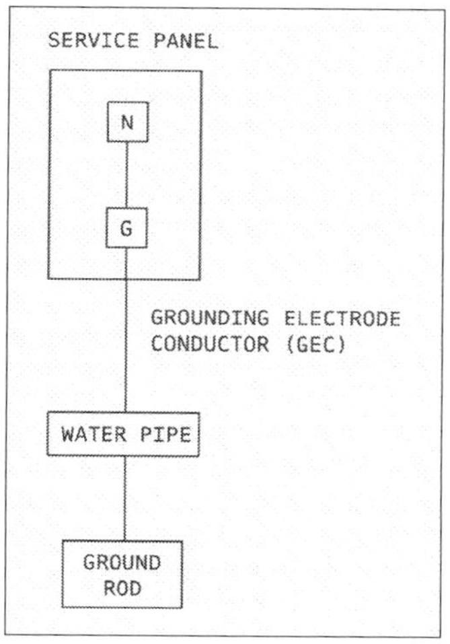

a) Identify any missing or incorrect connections
b) Label the minimum sizes required for each grounding conductor shown
c) Identify any bonding requirements that are not shown

SECTION 5: SHORT ANSWER AND PRACTICAL APPLICATION (15 Questions)

46. **Explain the difference between GFCI and AFCI protection, including:**

a) The hazards each is designed to protect against
b) How each functions to provide protection
c) Where each is required by the NEC

47. **A customer reports that half of their outlets in the living room have stopped working, while the other half still function normally. Outline your troubleshooting process, including:**

a) Possible causes
b) Testing procedures
c) Likely solutions

48. **Describe the proper procedure for sizing an overcurrent device and conductors for a motor circuit. Include considerations for:**

    a) Running overcurrent protection
    b) Branch-circuit short-circuit and ground-fault protection
    c) Conductor sizing for both circuit types

49. **Explain the requirements and procedure for bonding a swimming pool, including:**

    a) What components must be bonded
    b) Minimum size of bonding conductors
    c) Proper connection methods

50. **Detail the key differences between wiring methods permitted for:**

    a) Normal power circuits
    b) Emergency power circuits
    c) Critical care area circuits in hospitals

16

Full-Length Practice Exam Answer Key

SECTION 1: MULTIPLE-CHOICE QUESTIONS (1–20)

1. **C** — 24 inches required for direct-buried UF cable (NEC Table 300.5).
2. **B** — Derating starts when more than three conductors are in a conduit (NEC 310.15(C)(1)).
3. **C** — Continuous load requires conductor ampacity at 125% of the load (NEC 210.19(A)(1)).
4. **D** — Metal conduit is not considered a grounding electrode (NEC 250.52).
5. **B** — At least two 20-ampere small appliance branch circuits are required (NEC 210.11(C)(1)).
6. **C** — Bedrooms do not require GFCI protection (NEC 210.8(A)).
7. **C** — Maximum support distance for 1/2" EMT conduit is 10 feet (NEC 358.30(A)).
8. **A** — In a delta system, the line current equals the phase current.
9. **C** — Power factor = 1200 watts ÷ 1500 VA = **0.8**.
10. **B** — Resistance of a copper conductor increases with temperature.
11. **B** — Peak voltage = RMS × $\sqrt{2}$ = 120V × 1.414 = **169.7 volts**.
12. **B** — Line voltage = Phase voltage × $\sqrt{3}$ in a wye system.
13. **C** — Rigid Metal Conduit (RMC) provides best protection against physical damage.
14. **B** — A GFCI-protected receptacle is required under a kitchen sink (NEC 210.8(A)).

15. **B** — Special connectors listed for copper-to-aluminum connections must be used (NEC 110.14).

16. **B** — AFCI protection is required to protect against fire-causing arcs (NEC 210.12).

17. **C** — Maximum number of 90-degree bends between pull points is 360° total (NEC 358.26).

18. **B** — PPE category is based on available fault current and clearing time (NFPA 70E Table 130.7(C)(15)(a)).

19. **A** — A non-contact voltage tester alone is insufficient to verify de-energization (NFPA 70E 120.5).

20. **B** — Minimum approach distance for 480V equipment is 3 feet 6 inches (NFPA 70E Table 130.4(D)(a)).

SECTION 2: CALCULATIONS (21-32)

21. **Voltage Drop Calculation:**

* Formula: Vd = (2 × K × I × L) ÷ A
* Vd = (2 × 12.9 × 30A × 150ft) ÷ 10,380
* Vd ≈ 11.2 volts
* % Voltage drop = (11.2 ÷ 240) × 100 ≈ 4.67%
* **NEC Reference:** 210.19(A) Informational Note 4

22. **Motor Full Load Current Calculation:**

* FLC = (25hp × 746W) ÷ (√3 × 460V × 0.86 × 0.92)
* FLC ≈ 28 amperes
* Minimum conductor size: #10 AWG THWN copper
* OCPD: Sized per NEC 430.52

23. **Conduit Fill Calculation:**

* 6 × #10 THWN-2 (0.0211 in^2 each) + 4 × #12 THWN-2 (0.0133 in^2 each)
* Total area ≈ 0.1846 in^2
* EMT 1" allows up to 0.346 in^2 at 40% fill
* **Conclusion:** 1" EMT is sufficient.
* **NEC Reference:** Chapter 9, Tables 4 and 5

24. **Dwelling Load Calculation:**

* General lighting: 2,200 sqft × 3 VA = 6600 VA
* Small appliance: 2 × 1500 VA = 3000 VA
* Laundry: 1500 VA
* Range (demand factor): 12 kW × 80% = 9600 VA
* Dryer: 5500 VA
* Water heater: 4500 VA
* Dishwasher: 1200 VA
* A/C: 24A × 240V = 5760 VA
* **Total Load ≈ 33,660 VA**
* Load at 240V = 33,660 ÷ 240 ≈ 140.3A
* **Service Size:** 150A minimum (200A recommended for future capacity)
* **NEC Reference:** 220.82

25. **Box Fill Calculation:**

* 5 × #12 conductors @ 2.25 in^3 each = 11.25 in^3
* 3 × #10 conductors @ 2.5 in^3 each = 7.5 in^3
* 2 clamps @ 2.5 in^3
* 1 device @ 4.5 in^3 (twice largest conductor size)
* **Total volume needed:** 25.75 in^3
* 4-11/16" × 2-1/8" box provides ~42 in^3
* **Conclusion:** Box is sufficient.
* **NEC Reference:** 314.16

26. **Maximum Circuit Length for Voltage Drop (3% limit):**

* Formula: L = (Vd × A) ÷ (2 × K × I)
* L = (3.6V × 6530cm) ÷ (2 × 12.9 × 20A)
* L ≈ 45.6 feet
* **Conclusion:** About 45–50 feet maximum for 120V, 20A circuit.

27. **Three-Phase Feeder Calculation:**

a)

» Lighting load: 15kVA × 1.25 = 18.75kVA
» Receptacle load: 30kVA
» Motor Load : 20hp motor = 54A × 1.25 = 67.5A × 208V × √3 ≈ 24.3kVA
» Total load: ≈ 73kVA

» Load current: 73,000VA ÷ (208V × √3) ≈ 202A

b) **Feeder conductor size:** 250 kcmil Copper (255 A) conductor recommended.
c) 250 A breaker

28. **Load Balancing Calculation:**

* Phase A = 65A, Phase B = 47A, Phase C = 78A
* Average = (65 + 47 + 78) ÷ 3 ≈ 63.3A
* Maximum deviation = 78 - 63.3 = 14.7A
* % imbalance = (14.7 ÷ 63.3) × 100 ≈ 23.2%
* **Conclusion:** Above 20% — rebalance loads required.

29. **Short Circuit Current at Panel:**

* Transformer impedance Z = (480^2 × 5.75%) ÷ (500,000) = 0.0264 ohms
* Cable resistance = 0.0212 ohm × 2 × 200ft ÷ 1000 = 0.0085 ohms
* Total impedance Z_total = 0.0349 ohms
* Isc = (480V × √3) ÷ 0.0349 ≈ 23,800 amperes available fault current.

30. **Transformer Full Load Current:**

* Primary current = 75,000 ÷ (√3 × 480) ≈ 90.2A
* Secondary current = 75,000 ÷ (√3 × 208) ≈ 208A
* Overcurrent Protection Device (OCPD) size = 250A breaker per NEC 450.3(B).

SECTION 3: CODE REFERENCE QUESTIONS (10 Questions)

31. Dedicated Circuits for Dwelling Unit Appliances:

* NEC Article: 210.11(C)
* **Appliances requiring dedicated circuits:**
  » Refrigerator
  » Dishwasher
  » Microwave
* **Code References:**
  » Refrigerator: 210.52(B)(1) and 210.11(C)(1)
  » Dishwasher: 210.23(A)(2)
  » Microwave: 210.23(A)(2)

32. Receptacle Installations in Commercial Offices:

* NEC Article: 210.52(C), 210.63
* **Requirements:**
  - » General spacing: No point on a wall more than 6 feet horizontally from an outlet.
  - » Height: No specific height but usually standard installation (~18 inches AFF).
  - » GFCI: Required for receptacles serving sinks (NEC 210.8(B)).

33. Receptacle Installations in Patient care areas of healthcare facilities

NEC Article: 517.13, 517.18, 517.19

* **Requirements:**
* Receptacles per bed: 8 in general care, 14 in critical care.
* Emergency system: 4 (general) and 7 (critical) must be on essential power.
* Grounding: Dual path required—insulated copper + metal raceway/sheath.
* Routing: All grounding conductors must run with circuit conductors.
* Purpose: Ensure safety and reduce electrical shock hazards in patient areas.

34. Motor Disconnect Requirements:

* NEC Article: 430.102
* **Key Points:**
  - » Disconnect must be in sight from the motor and the controller.
  - » Controller can serve as disconnect if it meets NEC 430.109 requirements.
  - » Lockable disconnects allowed if out of sight.

35. Conductor Ampacity Adjustment Factors:

* NEC Article: 310.15(C)(1)
* **Example:**
  - » For more than 3 current-carrying conductors in a conduit:
    - › 4–6 conductors → 80% adjustment
    - › 7–9 conductors → 70% adjustment
* **Specific Example:**
  10 THWN-2 conductors → 50% derating required.

36. Working Space for Electrical Equipment (600V or Less):

* NEC Article: 110.26(A)
* **Key Requirements:**
    » Minimum depth: 3 feet (based on voltage-to-ground).
    » Minimum width: Width of the equipment or 30 inches, whichever is greater.
    » Minimum headroom: 6.5 feet.

37. Protection of Service Conductors:

* NEC Article: 230.91, 230.95
* **Key Points:**
    » Service conductors must have overcurrent protection.
    » Some exceptions for certain installations (like fire pumps).

38. Electrical Installations in Hazardous Locations:

* NEC Article: 500.5
* **Class I, Division 1 vs Division 2:**
    » Division 1: Explosive gases are normally present.
    » Division 2: Explosive gases are present only under abnormal conditions.

39. Emergency Systems Requirements:

* NEC Article: 700
* **Key Points:**
    » Separate wiring from normal systems.
    » Circuit protection and transfer equipment must meet specific standards.
    » Wiring methods must ensure survivability during emergencies.

40. Ground fault protection equipment

NEC Article: 230.95, 215.10, 210.13
Services: Ground-fault protection required for solidly grounded wye systems **over 150V to ground, 1000A or more, 1200A disconnect** (NEC 230.95).
Feeders: Same requirements apply to feeders supplying switchgear or panelboards downstream (NEC 215.10).
Branch Circuits: Not typically required unless specified for special applications (NEC 210.13).
Voltage threshold: Must exceed **150V to ground** and serve **3-phase, 4-wire wye systems**.

» **Purpose:** Protect equipment (not personnel) from damaging ground faults.

SECTION 4: DIAGRAM INTERPRETATION AND COMPLETION (41-45)

41. **Three-Way Switch Circuit:**

* **Missing Connections:**

   » Traveler wires between traveler terminals of both switches.
   » Power source to common terminal of first switch.
   » Load to common terminal of second switch.

Operation:
Switches alternate completing or breaking the circuit to control light from two locations.

42. **Electrical Riser Diagram for Three-Story Building:**

* **Violations Identified:**

   » Missing grounding electrode conductor.
   » No dedicated neutral for panel 3B.

* **Feeder Size for Panel 3B:**

   » Based on load calculation + 125% continuous load adjustment.

* **Grounding Connections:**

   » Proper bonding and grounding electrode connections must be completed.

43. **Single-Line Diagram for 480V Distribution:**

* **Overcurrent Protection:**

   » Main OCPD sized per service entrance rating.
   » Panel OCPD sized based on calculated loads.

* **Transformer Secondary Protection:**

   » Separate protection required within 10 feet (NEC 240.21(C)).

* **Grounding Points:**

   » Transformer secondary neutral grounded.

44. **Motor Control Circuit Diagram:**

* **Components:**

   » Start button (normally open)

» Stop button (normally closed)
» Contactor coil
» Overload relay

* **Operation:**

» Pressing Start energizes coil.
» Coil holds via auxiliary contact.
» Overload trips break the control circuit

45. **Grounding System Diagram for Service Entrance:**

* **Required Connections:**

» Main bonding jumper from neutral bar to enclosure.
» Grounding electrode conductor to electrode.

* **Minimum Sizes:**

» Based on NEC Table 250.66.

* **Bonding Requirements:**

» Bond water piping and structural steel.

SECTION 5: SHORT ANSWER AND PRACTICAL APPLICATION (46–60)

46–48. **Difference Between GFCI and AFCI Protection:**

* **GFCI:**

» Protects against electric shock (detects leakage to ground).
» Required in kitchens, bathrooms, outdoors, etc.

* **AFCI:**

» Protects against fire caused by arc faults.
» Required in bedrooms, living areas, laundry rooms.

* **NEC References:** 210.8, 210.12

49–51. **Troubleshooting Half-Power in Living Room:**

* **Possible Causes:**
  » Tripped breaker
  » Broken neutral
  » Loose receptacle backstab
* **Testing:**
  » Verify breaker status.
  » Voltage tests at receptacles.
  » Continuity testing of circuit.
* **Solutions:**
  » Repair broken neutral.
  » Tighten loose connections.

52–54. **Sizing Overcurrent Protection for Motors:**

* **Steps:**
  » Motor branch circuit short-circuit protection: 250% of FLC for breaker.
  » Overload protection: 115–125% of motor FLC.
  » Conductors sized based on 125% FLC.

55–57. **Bonding Swimming Pools:**

* **Components to Bond:**
  » Pool reinforcing steel.
  » Perimeter surfaces.
  » Equipment.
* **Bonding Conductor Size:**
  » Minimum 8 AWG copper.
* **Connection Methods:**
  » Listed clamps, exothermic welding.

58–60. **Wiring Methods for Different Systems:**

* **Normal Circuits:**
  » Standard NM or conduit wiring.
* **Emergency Systems:**
  » Separate conduits, specific survivability requirements.
* **Critical Care Areas:**
  » Redundant circuits, isolated grounding, metallic raceways.

Conclusion

The day I passed my journeyman exam, I walked out of the testing center feeling ten feet tall. I called Frank immediately, my voice cracking with excitement.

"Frank! I passed!"

There was a pause on the line, then his gravelly chuckle. "Well now, I'd have been mighty surprised if you hadn't."

We met at Donovan's later that evening, the same bar where three generations of electricians had celebrated their licenses. Frank ordered whiskey neat, I had a beer, and we sat at the scarred wooden bar where countless electricians had rested their elbows before us.

"You know," Frank said, swirling his glass, "that license is just a piece of paper. What matters is the electrician you've become while earning it."

I think about those words often, especially when I'm facing a tricky problem on a job site or training someone new to the trade. The exam tests what you know, but being a journeyman is about what you do with that knowledge.

I wrote this book because I remember the night before my exam, staring at a stack of code books, my head swimming with formulas and conductor sizes, and voltage drop calculations. I remember wishing I had someone sitting across the kitchen table from me, someone who'd been through it all before, who could tell me what really mattered.

If there's one last thing I can tell you, it's this: Trust what you know. The knowledge is in there, even when panic tries to convince you otherwise.

And when you pass—because you will—take a moment to think about all the hands that helped get you there. The journeymen who showed you tricks of the trade. The teachers who explained concepts until they clicked. The family that put up with your endless studying and shop talk.

Then find a bar like Donovan's, raise a glass, and get ready for the next chapter. The real work—the good work—is just beginning.

Bonus Resources

Ready to Ace Your Journeyman Electrician Exam?

Congratulations on reaching the end of this book - but your preparation doesn't stop here. We've created exclusive bonus tools to help you succeed:

Online Practice Tests — Simulate real exam conditions with randomized, timed tests
Digital Flashcards — Quickly memorize key facts and concepts anytime, anywhere
Audio Lessons — Reinforce your knowledge on the go

These resources are completely free for book owners and designed to maximize your confidence on exam day.

A. How to Access Your Bonus Materials:

Scan the QR code below with your mobile device:

Enter your unique access code from this book: journeyman2201

B. Stay Connected

We regularly update the platform with new resources - so bookmark the site and check back often!

C. Your Review Makes a Difference!

If our guide has helped you on your journey to becoming a licensed journeyman electrician, we'd be grateful if you'd share your experience.

Please take a moment to leave an honest review on Amazon.

Your feedback not only helps us improve our materials but also assists fellow electricians in finding reliable resources for their exam preparation.

List of Figures

List of Tables

References

1. Aresco, R. (2023, July 12). *Principle of electricity and DC & AC power supply: The basics.* Erie Institute of Technology. https://erieit.edu/electricity-dc-ac-power-supply-basics/
2. Austin. (2022, February 22). *The applications and limitations of Ohm's law.* My Tutor Source. https://ksa.mytutorsource.com/blog/ohms-law/
3. Awati, R. (2022, May). *What is conductor?* WhatIs.com. https://www.techtarget.com/whatis/definition/conductor
4. BBC. (n.d.). *Fuses and circuit breakers.* BBC Bitesize. https://www.bbc.co.uk/bitesize/guides/z2h4dxs/revision/3
5. *Conductors and insulators —KS3 Physics.* (n.d.). BBC Bitesize. https://www.bbc.co.uk/bitesize/articles/zd794xs
6. *Conductors, outlets, switches —understanding NSPIRE standards —1.* (2023). US Housing Consultants. https://nspire.us-hc.com/1/en/topic/electrical-conductor
7. Electric Circuit. (2018). In *Encyclopædia Britannica.* https://www.britannica.com/technology/electric-circuit
8. *Electrical formulas: basic concepts, units, and solved examples.* (2023). Testbook. https://testbook.com/physics-formulas/electrical-formulas
9. *Electrical switches and receptacles styles, types, and wiring.* (2020, June 9). Can Electric: Edmonton Electrical Contractor, Edmonton Electrician. https://canelectric.ca/blog/electrical-switches-and-receptacles-styles-types-and-wiring/
10. *Five practical applications of Ohm's law in daily life.* (2018, February 20). Ohm Law. https://ohmlaw.com/5-practical-applications-of-ohms-law-in-daily-life/
11. fxstageadmin. (2022, May 30). *Intro to load regulation for power supplies.* Advanced Conversion Technology. https://www.actpower.com/blog/load-regulation-power-supplies/
12. Gayathri. (2024, March 4). *Ohm's law, power sources, branches, nodes & loops.* CircuitBread; CircuitBread. https://www.circuitbread.com/tutorials/understanding-components-of-an-electrical-circuit
13. *Kirchhoff's laws and their applications.* (n.d.). Unacademy. https://unacademy.com/content/jee/study-material/physics/kirchhoffs-laws-and-their-applications/

14. Portal, E., & Parmar, J. (2013a, June 10). *How to calculate voltage regulation of distribution line.* EEP—Electrical Engineering Portal. https://electrical-engineering-portal.com/how-to-calculate-voltage-regulation-of-distribution-line

15. Portal, E., & Parmar, J. (2013b, June 10). *How to calculate voltage regulation of distribution line.* EEP—Electrical Engineering Portal. https://electrical-engineering-portal.com/how-to-calculate-voltage-regulation-of-distribution-line

16. *Receptacle and switch wiring.* (2018, December 22). Workforce LibreTexts. https://workforce.libretexts.org/Bookshelves/Construction/Building_Maintenance_and_Construction_-_Tools_and_Maintenance_Tasks_(Rutherford)/05%3A_Electrical_Systems/5.06%3A_Receptacle_and_Switch_Wiring

17. Tenny, K. M., & Keenaghan, M. (2023, May 22). *Ohm's Law.* PubMed; StatPearls Publishing. https://www.ncbi.nlm.nih.gov/books/NBK441875/

18. *The differences between thermal relays, fuses, and circuit breakers.* (2024). Asbeam. https://www.asbeam.com/news/breaker_relay_fuse_differences-cn.html

19. *The National Electrical Code (NEC).* (n.d.). Electrical Safety Foundation. https://www.esfi.org/workplace-safety/industry-codes-regulations/the-national-electrical-code-nec/

20. Turito. (2023, August 9). *Electrical components and their symbols.* US Learn. https://www.turito.com/learn/physics/electrical-components

21. *Understanding ohm's law: Basic fundamentals and applications.* (2023, October 25). Quarktwin Electronic; QuarkTwin. https://www.quarktwin.com/blogs/other/understanding-ohm-s-law-basic-fundamentals-and-applications/440?srsltid=AfmBOoqEWmoHbMfPh0DQKcrY_eaBeewVeEC6YYbHjltONpsOMlYAEBPw

22. *What are the basic principles of electricity?* (2018, September 19). Electrical Apprentice. https://electricalapprentice.co.uk/what-are-the-basic-principles-of-electricity/

23. Wright, G. (2022, April). *What is a circuit in electronics?* WhatIs.com. https://www.techtarget.com/whatis/definition/circuit

Made in United States
Orlando, FL
05 November 2026

90229884R00164